D1215269

Calculus
Without
Limits—*Almost*

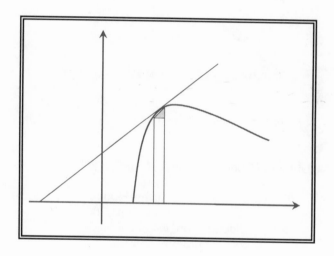

By John C. Sparks

Calculus Without Limits

Copyright © 2004, 2005, 2007
John C. Sparks

All rights reserved. No part of this book may be reproduced in any form—except for the inclusion of brief quotations in a review—without permission in writing from the author or publisher. The exceptions are all cited quotes, the poem "The Road Not Taken" by Robert Frost, the poem "Euclid Alone Has Looked on Beauty Bare" by Edna St. Vincent Millay (both poems appearing herein in their entirety), and the four geometric playing pieces that comprise Paul Curry's famous missing-area paradox.

Back cover photo by Curtis Sparks

ISBN: 1-4184-4124-4

First Published by Author House 9/10/2007

3rd Edition with Minor Enhancements and Corrections

Library of Congress Control Number: 2004106681

Published by **AuthorHouse**
1663 Liberty Drive, Suite 200
Bloomington, Indiana 47403
(800)839-8640
www.authorhouse.com

Produced by *Sparrow-Hawke* †*reasures*
Xenia, Ohio 45385

Printed in the United States of America

Dedication

I would like to dedicate <u>Calculus Without Limits</u>
To Carolyn Sparks, my wife, lover, and partner for 39 years;
And to Robert Sparks, American warrior, and elder son of geek;
And to Curtis Sparks, reviewer, critic, and younger son of geek;
And to Roscoe C. Sparks, deceased, father of geek.

From Earth with Love

Do you remember, as do I,
When Neil walked, as so did we,
On a calm and sun-lit sea
One July, Tranquillity,
Filled with dreams and futures?

> For in that month of long ago,
> Lofty visions raptured all
> Moonstruck with that starry call
> From life beyond this earthen ball...
> Not wedded to its surface.

But marriage is of dust to dust
Where seasoned limbs reclaim the ground
Though passing thoughts still fly around
Supernal realms never found
On the planet of our birth.

> And I, a man, love you true,
> Love as God had made it so,
> Not angel rust when then aglow,
> But coupled here, now rib to soul,
> Dear Carolyn of mine.

July 2002: 33rd Wedding Anniversary

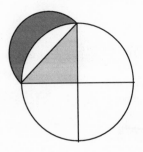

Hippocrates' Lune: Circa 440BCE

This is the earliest known geometric figure with two curvilinear boundaries for which a planar area could be exactly determined.

Forward

I first began to suspect there was something special about John Sparks as a teacher back in 1994 when I assumed the role of department chair and got a chance to see the outstanding evaluations he consistently received from his students. Of course I knew that high student ratings don't always equate to good teaching. But as I got to know John better I observed his unsurpassed enthusiasm, his unmitigated optimism and sense of humor, and his freshness and sense of creativity, all important qualities of good teaching. Then when I attended several seminars and colloquia at which he spoke, on topics as diverse as *Tornado Safety, Attention Deficit Syndrome* and *Design of Experiments,* I found that his interests were wide-ranging and that he could present material in a clear, organized and engaging manner. These are also important qualities of good teaching. Next I encountered John Sparks the poet. From the poems of faith and patriotism which he writes, and the emails he periodically sends to friends, and the book of poems, *Mixed Images,* which he published in 2000, I soon discovered that this engineer by trade is a man with one foot planted firmly on each side of the intellectual divide between the arts and the sciences. Such breadth of interest and ability is most assuredly an invaluable component of good teaching. Now that he has published *Calculus without Limits,* the rest (or at least more) of what makes John Sparks special as a teacher has become clear. He has the ability to break through those aspects of mathematics that some find tedious and boring and reveal what is fascinating and interesting to students and what engages them in the pursuit of mathematical knowledge. By taking a fresh look at old ideas, he is able to expose the motivating principles, the intriguing mysteries, the very guts of the matter that are at the heart of mankind's, and especially this author's, abiding love affair with mathematics. He manages to crack the often times opaque shell of rules and formulas and algorithms to bring to light the inner beauty of mathematics. Perhaps this completes my understanding of what is special about John Sparks as a teacher. Or perhaps he still has more surprises in store for me. Anyway, read this book and you will begin to see what I mean.

Al Giambrone
Chairman
Department of Mathematics
Sinclair Community College
Dayton, Ohio   October 2003

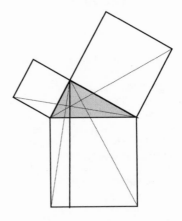

Euclid's Windmill: Circa 300 BCE

*'Windmill diagram' associated with
Euclid's proof of the Pythagorean Theorem:
The Elements, Book 1, Proposition 47*

Euclid Alone Has Looked on Beauty Bare

Euclid alone has looked on Beauty bare.
Let all who prate of Beauty hold their peace,
And lay them prone upon the earth and cease
To ponder on themselves, the while they stare
At nothing, intricately drawn nowhere
In shapes of shifting lineage; let geese
Gabble and hiss, but heroes seek release
From dusty bondage into luminous air.

> O blinding hour, O holy, terrible day,
> When first the shaft into his vision shown
> Of light anatomized! Euclid alone
> Has looked on Beauty bare. Fortunate they
> Who, though once and then but far away,
> Have heard her massive sandal set on stone.

Edna St. Vincent Millay

Table of Contents

Table of Contents cont

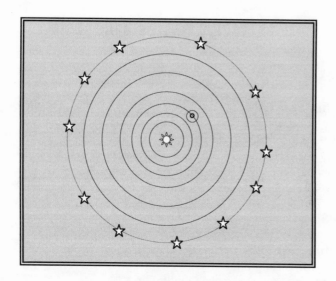

The Copernican Heliocentric Solar System

*The emerging and evolving cosmological model when
Sir Isaac Newton was born*

Significance

The wisp in my glass on a clear winter's night
Is home for a billion wee glimmers of light,
Each crystal itself one faraway dream
With faraway worlds surrounding its gleam.

And locked in the realm of each tiny sphere
Is all that is met through an eye or an ear;
Too, all that is felt by a hand or our love,
For we are but whits in the sea seen above.

Such scales immense make wonder abound
And make a lone knee touch the cold ground.
For what is this man that he should be made
To sing to The One Whose breath heavens laid?

July 1999

List of Tables, Figures, and Poems

Tables

Figures

Figures...cont

Figures...cont

Figures...cont

Poems

1) Introduction

"If it was good enough for old Newton,
It is good enough for me."
Unknown

1.1) General

I love calculus! This love affair has been going on since the winter of 1966 and, perhaps a little bit before. Indeed, I remember purchasing my first calculus textbook (by Fobes and Smyth) in December of 1965 and subsequently pouring through its pages, pondering the meaning of the new and mysterious symbols before me. Soon afterwards, I would be forever hooked and yoked as a student, teaching assistant, teacher, and lifelong admirer.

Over the years, my rose-colored perspective has changed. I have discovered like many other instructors that most students don't share an "aficionado's" enthusiasm for calculus (as we do). The reasons are many, ranging from attitude to aptitude, where a history of substandard "classroom-demonstrated" mathematical aptitude can lead to poor attitude. The tragedy is that with some students the aptitude is really there, but it has been covered over with an attitude years in the making that says, "I just can't do mathematics." These students are the target audience for this book. A long-simmering mathematical aptitude, finally discovered and unleashed, is a marvelous thing to behold, which happens to be my personal story.

So what has happened to calculus over the last four decades in that it increasingly seems to grind students to dust? Most textbooks are absolutely beautiful (and very expensive) with articles and items that are colored-coded, cross-referenced, and cross-linked. Additionally, hand-done "engineering drawings" have been replaced by magnificent 3-D computer graphics where the geometric perspective is absolutely breathtaking and leaves little to the imagination. *Note: I have to confess to a little jealously having cut my teeth on old fashion black-and-white print augmented with a few sketches looking more like nineteenth-century woodcuts.*

The answer to the above question is very complex, more complex (I believe) than any one person can fathom. Let it suffice to say that times have changed since 1965; and, for students today, time is filled with competing things and problems that we baby boomers were clueless about when of similar age. Much of this is totally out of our control.

So, what can we control? In our writing and explanation, we can try to elucidate our subject as much as humanly possible. I once heard it said by a non-engineer that an engineer is a person who gets excited about boring things. Not true! As an engineer and educator myself, I can tell you that an engineer is a person who gets excited about very exciting things—good things of themselves that permeate every nook and cranny of our modern American culture. The problem as the warden in the Paul Newman movie <u>Cool Hand Luke</u> so eloquently stated, "is a failure to communicate." The volume in your hands, <u>Calculus Without Limits</u>, is a modern attempt to do just that—communicate! Via a moderate sum of pages, my hope is that the basic ideas and techniques of calculus will get firmly transferred to a new generation, ideas and techniques many have called the greatest achievement of Western science.

The way this book differs from an ordinary "encyclopedic-style" textbook is twofold. One, it is much shorter since we cover only those ideas that are central to an understanding of the **calculus of *a* real-valued function of a single real-variable**. *Note: Please don't get scared by the last bolded expression and run off. You will understand its full meaning by the end of Chapter 4.* The shortness is also due to a lack of hundreds upon hundreds of skill-building exercises—very necessary if one wants to become totally competent in a new area of learning. However, a minimal set of exercises (about 230 in all) is provided to insure that the reader can verify understanding through doing. Two, as stated by the title, this is a calculus book that minimizes its logical dependence on the limit concept (*Again, Chapter 4.*). From my own teaching experience and from reading book reviews on web sites, **the limit concept seems to be the major stumbling block preventing a mastery of engineering-level calculus**. The sad thing is that it doesn't need to be this way since calculus thrived quite well without limits for about 150 years after its inception; relying instead on the differential approach of Newton and Leibniz.

Differentials—little things that make big ideas possible—are the primary means by which calculus is developed in a book whose title is <u>Calculus Without Limits</u>. The subtitle —Almost refers to the fact that the book is not entirely without limits. Section 4.3 provides an intuitive and modern explanation of the limit concept. From that starting point, limits are used thereafter in a handful (quite literally) of key arguments throughout the book .

Now for the bad news! One, <u>Calculus Without Limits</u> is a primer. This means that we are driving through the key ideas with very few embellishments or side trips. Many of these embellishments and side trips are absolutely necessary if one wants a full understanding of all the technical power available in the discipline called calculus. To achieve full mastery, nothing takes the place of all those hours of hard work put into a standard calculus sequence as offered through a local college or university. This book should be viewed only as an aid to full mastery—a starter kit if you will. Two, <u>Calculus Without Limits</u> is not for dummies, morons, or anyone of the sort. <u>Calculus Without Limits</u> is for those persons who want to learn a new discipline and are willing to take the time and effort to do so, provided the discipline is presented in such a manner as to make in-depth understanding happen. If you don't want to meet <u>Calculus Without Limits</u> halfway—providing your own intellectual work to understand what is already written on each page—then my suggestion is to leave it on the book-seller's shelf and save yourself some money.

1.2) Formats, Symbols, and Book Use

One of my interests is poetry, having written and studied poetry for over a decade. Several theme-supporting examples of my own poetry appear in this book. I have also included "The Road Not Taken" by Robert Frost and "Euclid Alone Has Looked on Beauty Bare" by Edna St. Vincent Millay.

If you pick up a textbook on poetry and thumb the pages, you will see poems interspersed between explanations, explanations that English professors will call prose. Prose differs from poetry in that it is a major subcategory of how language is used.

Prose encompasses all the normal uses: novels, texts, newspapers, magazines, letter writing, and such. But poetry is different! Poetry is a highly charged telescopic (and sometimes rhythmic) use of the English language, which is employed to simultaneously convey a holographic (actual plus emotional) description of an idea or an event. Poetry not only informs our intellect, it infuses our soul. Poetry's power lies in the ability to do both in a way that it is easily remembered. Poetry also relies heavily on concision: not a word is wasted! Via the attribute of concision, most poetry when compared to normal everyday prose looks different Thus, when seen in a text, poems are immediately read and assimilated differently than the surrounding prose. *Note: The font used for poetry and quotes in Calculus without Limits is Georgia as shown here. Arial is used for prose. Impact is used for Major Titles.*

So what does poetry have to do with mathematics? Any mathematics text can be likened to a poetry text. In it, the author is interspersing two languages: a language of qualification (English in the case of this book) and a language of quantification (the universal language of algebra). The way these two languages are interspersed is very similar to that of the poetry text. When we are describing, we use English prose interspersed with an illustrative phrase or two of algebra. When it is time to do an extensive derivation or problem-solving activity—using the highly-changed, dripping-with-mathematical-meaning, and concise algebraic language—then the whole page (or two or three pages!) may consist of nothing but algebra. Algebra then becomes the alternate language of choice used to unfold the idea or solution. Calculus Without Limits and without apology follows this general pattern, which is illustrated in the next paragraph by a discussion of the quadratic formula.

☺

Let $ax^2 + bx + c = 0$ be a quadratic equation written in the standard form as shown with $a \neq 0$. Then $ax^2 + bx + c = 0$ has two solutions (including complex and multiple) given by the formula highlighted below, called the quadratic formula.

$$x = \frac{-b \pm \sqrt{b^2 - 4ac}}{2a}.$$

To solve a quadratic equation, using the quadratic formula, one needs to apply the following four steps considered to be a *solution process*.

1. Rewrite the quadratic equation in standard form.
2. Identify the two coefficients and constant term $a, b, \& c$.
3. Apply the formula and solve for the two x values.
4. Check your two answers in the original equation.

To illustrate this four-step process, we will solve the quadratic equation $2x^2 = 13x + 7$.

$$\overset{1}{\mapsto}: 2x^2 = 13x + 7 \Rightarrow$$
$$2x^2 - 13x - 7 = 0$$
****

$$\overset{2}{\mapsto}: a = 2, b = -13, c = -7$$
****

$$\overset{3}{\mapsto}: x = \frac{-(-13) \pm \sqrt{(-13)^2 - 4(2)(-7)}}{2(2)} \Rightarrow$$

$$x = \frac{13 \pm \sqrt{169 + 56}}{4} \Rightarrow$$

$$x = \frac{13 \pm \sqrt{225}}{4} = \frac{13 \pm 15}{4} \Rightarrow$$

$$x \in \{-\tfrac{1}{2}, 7\}$$
****

$$\overset{4}{\mapsto}: \text{This step is left to the reader.}$$

☺

Taking a look at the text between the two happy-face symbols☺ ☺, we first see the usual mixture of algebra and prose common to math texts. The quadratic formula itself, being a major algebraic result, is highlighted in a shaded double-bordered (SDB) box.

We will continue the use of the SDB box throughout the book, highlighting all major results—and warnings on occasion! If a process, such as solving a quadratic equation, is best described by a sequence of enumerated steps, the steps will be presented in indented, enumerated fashion as shown. Not all mathematical processes are best described this way, such as the process for solving any sort of word problem. The reader will find both enumerated and non-enumerated process descriptions in <u>Calculus Without Limits</u>. The little bit of *italicized text* identifying the four steps as a solution process is done to cue the reader to a very important thought, definition, etc. Italics are great for small phrases or two-to-three word thoughts. The other method for doing this is to simply insert the whole concept or step-wise process into a SDB box. *Note: italicized 9-font text is also used throughout the book to convey special cautionary notes to the reader, items of historical or personal interest, etc.* Rather than footnote these items, I have chosen to place them within the text exactly at the place where they augment the overall discussion.

Examining the solution process proper, notice how the solution stream lays out on the page much like poetry. The entire solution stream is indented; and each of the four steps of the solution process is separated by four asterisks ****, which could be likened to a stanza break. If a solution process has not been previously explained and enumerated in stepwise fashion, the asterisks are omitted. The new symbol $\overset{1}{\mapsto}:$ can be roughly translated as "The first step proceeds as follows." Similar statements apply to $\overset{2}{\mapsto}:$ $\overset{3}{\mapsto}:$ and $\overset{4}{\mapsto}:$. The symbol $\Rightarrow$ is the normal "implies" symbol and is translated "This step implies (or leads to) the step that follows". The difference between "$\overset{1}{\mapsto}:$" and "$\Rightarrow$" is that $\overset{1}{\mapsto}:$ is used for major subdivisions of the solution process (either explicitly referenced or implied) whereas $\Rightarrow$ is reserved for the stepwise logical implications within a single major subdivision.

Additionally, notice in our how-to-read-the-text example that the standard set-inclusion notation $\in$ is used to describe membership in a solution set. This is true throughout the book.

Other standard set notations used are: union $\cup$, intersection $\cap$, existence $\exists$, closed interval $[a,b]$, open interval (a,b), half-open-half-closed interval $(a,b]$, not a member $\notin$, etc. The symbol $\therefore$ is used to conclude a major logical development; on the contrary, $\therefore$ *is not used* to conclude a routine problem.

Though not found in the quadratic example, the usage of the infinity sign ∞ is also standard. When used with interval notation such as in $(-\infty, b]$, minus infinity would denote a semi-infinite interval stretching the negative extent of the real number starting at and including b (since $]$ denotes closure on the right). Throughout the book, all calculus notation conforms to standard conventions—although, as you will soon see, not necessarily standard interpretations. Wherever a totally new notation is introduced (which is not very often), it is explained at that point in the book—following modern day "just-in-time" practice.

Lastly, in regard to notation, I would like you to meet

The Happy Integral

$$\int_a^b \overset{\cdot\cdot}{\cup} dx$$

The happy integral is used to denote section, chapter, and book endings starting in Chapter 2. *One* happy integral denotes the end of a section; *two* happy integrals denote the end of a chapter; and *three* happy integrals denote the end of the main part of the book. Subsections (not all sections are sub-sectioned—just the longer ones) are not ended with happy integrals.

Note: you will find out about real integrals denoted by the' foreboding and esoteric-looking' symbol $\int_a^b f(x)dx$ *starting in Chapter 6. In the meantime, whenever you encounter a happy integral, just be happy that you finished that much of the book!*

Calculus without Limits is suitable for either self study (recommended use) or a one-quarter introductory calculus course of the type taught to business or economic students. The book can also be used to supplement a more-rigorous calculus curriculum. As always, there are many ways a creative mathematics instructor can personalize the use of available resources. The syllabus below represents one such usage of Calculus without Limits as a primary text for an eleven-week course of instruction.

Suggested Syllabus for Calculus without Limits: Eleven-Week Instructor-led Course			
Week	**Chapter**	**Content**	**Test**
1	1, 2, 3	Introduction, Barrow's Diagram, *Two Fundamental Problems*	
2	4.1-4.4	Functions & Inverse Functions	1, 2, 3
3	4.5-4.8	Slopes, Change Ratios and Differentials	
4	5.1-5.4	*Solving the First Problem*, Derivatives and Applications	
5	5.5-5.7	Higher Order Derivatives and Advanced Applications	4, 5
7	6.1-6.4	Antiprocesses, Antidifferentiation and Basic Applications	
8	7.1-7.3	*Solving the Second Problem*, Continuous Sums, Definite Integral, Fundamental Theorem	
9	7.4-8.1	Geometric Applications, Intro to Differential Equations	6, 7
10	8.2	Differential Equations in Physics	
11	8.3, 11	Differential Equations in Finance, Conclusions, and Challenge	Final
Note 1: All primary chapters (3 through 10) and most of the sections within these chapters have associated exercises. It is recommended that the instructor assign all exercises appearing in the book. A complete set of answers starts on page 360.			
Note 2: The student is encouraged to make use of the ample white space provided in the book for the hand-writing of personalized clarifications and study notes.			

Table 1.1 Calculus without Limits Syllabus

1.3) Credits

No book such as this is an individual effort. Many people have inspired it: from concept to completion. Likewise, many people have *made it so* from drafting to publishing. I shall list just a few and their contributions.

Silvanus Thompson, I never knew you except through your words in <u>Calculus Made Easy</u>; but thank you for propelling me to fashion an every-person's update suitable for a new millennium. Melcher Fobes, I never knew you either except for your words in <u>Calculus and Analytic Geometry</u>; but thank you for a calculus text that sought—through the power of persuasive prose combined with the language of algebra—to inform and instruct a young student—then age 18. Books and authors such as these are a rarity—definitely out-of-the-box!

To those great Americans of my youth—President John F. Kennedy, John Glenn, Neil Armstrong, and the like—thank you all for inspiring an entire generation to think and dream of bigger things than themselves. This whole growing-up experience was made even more poignant by the fact that I am a native Ohioan, a lifelong resident of the Dayton area (home of the Wright Brothers).

To my many readers, thank you all for burning through the manuscript and refining the metal. To Dr. Som Soni and Mr. Vincent Miller, thank you both for painstakingly combing the first edition for errors and making the subsequent editions possible.

To my two editors, Curtis and Stephanie Sparks, thank you for helping the raw material achieve full publication. This has truly been a family affair.

To my wife Carolyn, the Heart of it All, what can I say. You have been my constant and loving partner for some 39 years now. You gave me the space to complete this project and rejoiced with me in its completion. As always, we are a proud team!

John C. Sparks
February 2004, January 2005
Xenia, Ohio

1.3) Changes in the 2007 Edition

It has been four years since I first published <u>Calculus Without Limits</u>, and, as with any modern product, one strives towards continuous improvement of the same. Having worked in organizational quality-improvement for over a decade, I have learned that all major processes and products should be re-examined on a regular basis for improvement opportunities—hence, the 3rd Edition of <u>Calculus Without Limits</u>.

The one major change is the addition of a subject-matter **Index**. Minor changes include an Epilogue (a reflection of sorts on America's need to remain preeminent in science and technology), a demonstration of the Pythagorean Theorem using calculus (Section 5.4), and the use of the derivative in solving a heat transfer problem (also Section 5.4). Where appropriate, several new diagrams accompany the changes, which make <u>Calculus Without Limits</u> very much a diagram-rich work!

Finally, all original arguments have been given the 'sanity check' one more time. In a few cases, they were found lacking; and revised arguments have been introduced. Having a book on the open market for three or four years is a great way to discover the dross throughout the time-honored process of verbal feedback. Thanks again to all my readers!

John C. Sparks
October 2007
Xenia, Ohio

A Season for Calculus

Late August
Brings an end to limits,
Chained derivatives,
Constraints—optimized and otherwise—
Boundary conditions,
Areas by integrals,
And long summer evenings.
My equally fettered students,
Who moaned continuously
While under tight mathematical bondage,
Will finally be released—
Most with a pen-stroke of mercy!
Understandably,
For meandering heads just barely awake,
Newton's infinitesimal brainchild
Presented no competition when pitted against
Imagined pleasures faraway,
And outside
My basement classroom.
Always the case...

But, there are some,
Invariably a few,
Who will see a world of potential
In one projected equation
And opportunities stirring
In the clarifying scribble...

August 2001

Teaching mathematics—a bittersweet occupation

2) Barrow's Diagram

Calculus is ranked as one of the supreme triumphs of Western science. Current equivalents include the first manned lunar landing in 1969 or the decoding of the human genome in 2000. *Note: My personal lifespan has witnessed both the advent and continuing cultural fallout from each of these aforementioned equivalents.* Like most modern-day technical achievements, calculus has taken many minds to develop. Granted, these minds have not operated in the context of a highly organized team with intricately interlaced functions as in the two examples mentioned. Nonetheless, these inquisitive, capable minds still examined and expanded the ideas of their intellectual predecessors through the course of almost two millennia (though a *Western* intellectual hiatus occupied much of this time interval).

A mathematician can almost envision these minds interacting and enhancing each other via **Figure 2.1**, which has embedded within it a graphic mini-history of calculus.

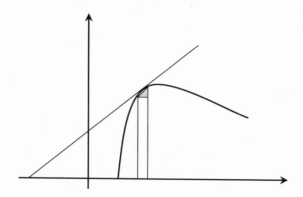

Figure 2.1 Barrow's Diagram

Figure 2.1 was originally created by Isaac Barrow (1630-1677) who was a geometer, first holder of the Lucasian chair at Cambridge, and a teacher/mentor to Sir Isaac Newton. Even today, you will see bits and pieces of Barrow's diagram, perhaps its entirety, used in any standard calculus text.

Barrow's 350-year-old diagram is proof that a powerful idea conveyed by a powerful diagrammatic means never dies. In this chapter, we will reflect upon his diagram as a creative masterpiece, much like a stained-glass window or painting.

Table 2.1 is an *artist* guide to Barrow's diagram, linking selected mathematicians to seven coded features. The guide is not meant to be complete or exhaustive, but does illustrate the extent of mathematical *cross-fertilization* over the course of two millennia.

Name	Coded Diagram Feature						
	T	**L**	**R**	**C**	**A**	**XY**	**IDT**
Pythagoras 540 BC	X	X	X				
Archimedes 287-212 BC	X	X	X	X	X		
Descartes 1596-1650		X		X		X	
Barrow 1630-1677	X	X	X	X	X	X	
Newton 1642-1727	X	X	X	X	X		
Leibniz 1646-1716	X	X	X	X	X		
Gauss 1777-1855	X	X	X	X	X	X	X
Cauchy 1789-1857	X	X	X	X	X	X	X
Riemann 1826-1866	X	X	X	X	X	X	X
Code	**Feature Description**						
T	Small shaded right triangle						
L	Straight line						
R	Tall slender rectangle						
C	Planar curve						
A	Area between the curve and triangle						
XY	Rectangular coordinate system						
IDT	In-depth theory behind the diagram						

Table 2.1 Guide to Barrow's Diagram

More will be said about these mathematicians and their achievements in subsequent chapters. But for the moment, I want you to pause, reflect upon the past, and just admire **Figure 2.1** as you would a fine painting. When finished, take a stroll down to **Figure 2.2** and do the same. **Figure 2.2** (non-annotated version) is the drawing used by Archimedes to perform his famous parabolic quadrature. In this result, Archimedes showed the area enclosed by the parabola sector equals two-thirds of the area enclosed by the gray rectangle and four-thirds of the area enclosed by the black triangle. The two assumptions are that 1) all three figures have common bases and 2) the apex of the triangle corresponds to the greatest extent of the parabolic arc from the common base. Imagine Barrow and Newton pondering the same figure while in search of new mathematics from the old.

Figure 2.2 Archimedean Parabolic Quadrature

Euclid's Beauty Revisited

Never did Euclid, as Newton ', discern
Areas between an edge and a curve
Where ancient precisions of straight defer
To infinitesimal addends of turn,
Precisely tallied in order to learn
Those planes that Euclid could only observe
As beauty...then barren of quadrature
And numbers for which Fair Order did yearn.

Thus beauty of worth meant beauty in square,
Or, those simple forms that covered the same.
And, though, Archimedes reckoned with care
Arcs of exhaustion no purist would claim,
Yet, his were the means for Newton to bare
True Beauty posed...in Principia's frame.

October 2006

We close Chapter 2 with **Figure 2.3** that presents two different non-algebraic visual proofs of the Pythagorean Theorem using constructions attributed to Pythagoras himself. *All six white-shaded right triangles are of identical size.* Armed with this simple fact, can you *truly see* the sum-of-squares Pythagorean identity once all white-shaded triangles are removed or 'taken away'?

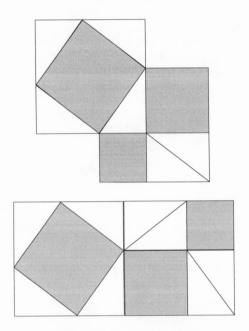

**Figure 2.3: Two Visual Proofs
Of the Pythagorean Theorem Using
Traditional Constructions**

$$\int_a^b \overset{\bullet\bullet}{\cup} dx \; \int_a^b \overset{\bullet\bullet}{\cup} dx$$

3) The Two Fundamental Problems of Calculus

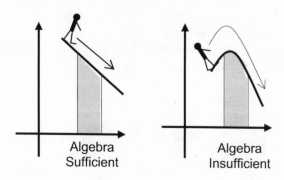

| Algebra Sufficient | Algebra Insufficient |

Figure 3.1: Two Paths of Varying Complexity

In **Figure 3.1**, the little stick person (a regular feature throughout the book) walks twice in the direction indicated by the arrows on two separate paths. The differences between these two paths are quite profound and distinguish *mere* algebra from calculus. Hence, we will call the leftmost path *the path of algebra* and the rightmost path *the path of calculus*.

On the path of algebra, our stick figure walks atop a line segment. Although no numbers are given, we have an intuitive sense that the *slope* associated with this walk is *always constant* and *always negative*. From algebra, if (x_1, y_1) and (x_2, y_2) are *any* two distinct points lying on a line segment, then the slope m of the associated line segment is defined by the well-recognized straight-line slope formula

$$m = \frac{y_2 - y_1}{x_2 - x_1}.$$

This definition implies that *any* two distinct points, no matter how close or far apart, can be used to calculate m as long as both points are directly located on the given line segment.

The definition of slope also implies that if these calculations are done correctly then m will be the same *or constant* for every pair of distinct points (see **Figure 3.2**) we chose, again, as long as both points are directly located on the line segment. Hence, the use of the straight-line slope formula substantiates our intuition that m is *always constant*. Secondly, there is an intuitive sense that m is *always negative* since our stick figure steadily decreases in elevation as the walk proceeds in the direction of the arrow. Again, the straight-line slope formula can be easily used to substantiate our intuition.

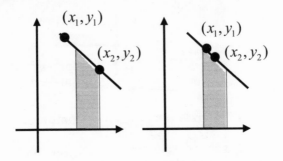

Figure 3.2: Different Points, Same Slope

Next, consider the shaded planar area below the path of algebra. This area, which we will call A, is enclosed by the horizontal axis, the two vertical line segments, and the sloping line segment. How would one calculate this area? Since the shaded area is a trapezoid, the answer is given by the associated trapezoidal area formula

$$A = \tfrac{1}{2}(b + B)h,$$

where B and b are the lengths of the two vertical lines segments and h is the horizontal distance between them. Notice that the shaded area A has four linear—or *constant sloping*—borders. This fact makes a simple formula for A possible. *In summary, when on the path of algebra, elementary algebraic formulas are sufficient to calculate both the slope of the line segment and the area lying underneath the line segment.*

Turning now to the path of calculus, our stick person walks atop a curve. During this walk, there is an intuitive sense that the slope is *always changing* as the person travels from left to right. At the start of the walk, our figure experiences a positive slope; at the end of the walk, a negative slope. And, somewhere in between, it looks like our figure experiences level ground—a high point where the slope is zero! So, how does one compute the slope along a curve where the slope is *always changing*? In particular, how does one compute the slope for a specific point, P, lying on the curve as shown in **Figure 3.3**? Perhaps we could start by enlisting the aid of the straight-line slope formula. But, the question immediately becomes, which two points (x_1, y_1) and (x_2, y_2) on the curve should we use? Intuition might say, choose two points close to P. But, is this exact? Additionally, suppose P is a point near or on the hilltop. Two points close together and straddling the hilltop could generate either a positive or negative slope—depending on the relative y_1 and y_2 values. Which is it? Obviously, the straight-line slope formula is insufficient to answer the question, what is the slope when a line is replaced by a curve? In order to answer this basic question, more powerful and more general slope-generating techniques are needed.

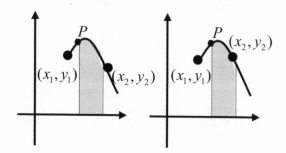

Figure 3.3: Slope Confusion

The First Fundamental Problem of Calculus:

Find the exact slope for any point P
Located on a general curve

Turning our attention to the shaded region under the curve, we are compelled to ask, what is the area? Immediately, the upper curved boundary presents a problem. It is not a border for a triangle, square, trapezoid, or rectangle. If it were (where each of the aforementioned figures has a *constant-sloping* boundary) then we might be able to use a standard area formula—or combination thereof—to produce an exact answer. No such luck. We can go ahead and approximate the shaded area, but our approximation will be non-exact and subject to visual error—the same error problem we had when trying to use the straight-line slope formula to find slopes for a curve.

The Second Fundamental Problem of Calculus:

Find the exact area for a planar region
Where at least part of the boundary is a general curve

This book is about solving the first and second fundamental problems of calculus. In the course of doing so, a marvelous set of mathematical tools will be developed that will greatly enhance your mathematical capability. The tools developed will allow you to solve problems that are simply unanswerable by algebra alone. Typically, these are problems where general non-linear curves prohibit the formulation of simple algebraic solutions when trying to find geometric quantities—quantities that can represent just about every conceivable phenomenon under the sun. Welcome to the world of calculus.

Note: The reader might ask the question, why isn't the circle included in the list of elementary areas above, a figure with a curved or non-constant-sloping boundary and a figure for which we have an elementary area formula?

The formula $A = \pi r^2$ was derived over 2200 years ago by Archimedes, using his exhaustion method, a rudimentary form of calculus. Though simple in algebraic appearance, sophisticated mathematical methods not to be seen again until the early European Renaissance were required for its development.

$$\int_a^b \smile dx \quad \int_a^b \smile dx$$

Chapter Exercises

We end Chapter 3 with two exercises that cleverly illustrate the visual geometric limitations of the human eye. Indeed, the eye alone without the supporting benefit of precise measurement can be deceiving. Transferring this observation to mathematics, one could say mathematics as a discipline needs to harmonize two sources of knowledge—playful intuition and precise logic—called left-right-brain integration in modern terms. The two exercises deliver both puzzle-solving fun and necessary practice for building the critical skill of knowledge harmonization.

1. Paul Curry was an amateur magician who practiced his trade in New York City. In 1953, Curry developed the now widely-known Curry's Paradox, as shown in **Figure 3.4,** where the two polygons are created from two identical sets of geometric playing pieces containing four areas each. The challenge is obvious, how did the square hole get into the upper polygon?

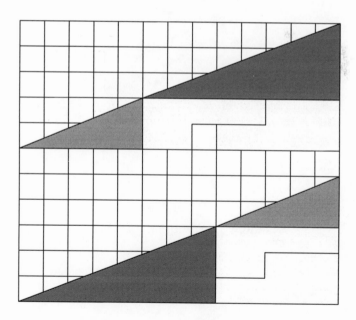

Figure 3.4: Curry's Paradox

2. Two identical Tangram sets consisting of the seven traditional playing pieces are used to construct both quadrilaterals as shown in **Figure 3.5**. The challenge is to offer a rational explanation for the disappearance of the irregular-shaped hole.

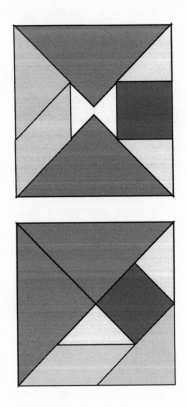

Figure 3.5: Tangram Paradox

4) Foundations

*"If I have seen further, it is by
Standing on the shoulders of Giants." Sir Isaac Newton*

4.1) Functions: Input to Output

The mathematical concept called a *function* is foundational to the study of calculus. Simply put: *To have calculus, we first must have a function*. With this statement in mind, let's define in a practical sense what is meant by the word *function*.

<u>Definition</u>: a *function* is any process where numerical input is transformed into numerical output with the operating restriction that each unique input must lead to one and only one output.

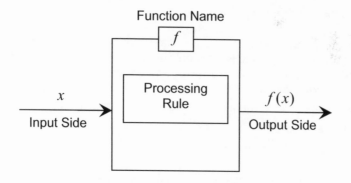

Figure 4.1: The General Function Process

Figure 4.1 is a diagram of the general function process for a function named f. Function names are usually lower-case letters chosen from f, g, h, etc.

When a mathematician says, let f be a function, the entire input-output process—start to finish—comes into discussion. If two different function names are being used in one discussion, then two different functions are being discussed, often in terms of their relationship to each other. The variable x (see **Appendix A** for a discussion of the true meaning of x) is the *independent or input* variable; it is independent because any specific input value can be freely chosen. Once a specific input value is chosen, the function then processes the input value via the processing rule in order to create the *output variable $f(x)$*. The *output variable $f(x)$* is also called the *dependent variable* since its value is entirely determined by the action of the processing rule upon x. Notice that the complex symbol $f(x)$ reinforces the fact that output values are created by direct action of the function process f upon the independent variable x. Sometimes, a simple y will be used to represent the output variable $f(x)$ when it is well understood that a function process is indeed in place.

Three more definitions are important when discussing functions. In this book, we will only study **real-valued functions of a real variable; these are functions where both the input and output variable must be a real number.** The set of all possible input values for a function f is called the *domain* and is denoted by the symbol Df. The set of all possible output values is called the *range* and is denoted by Rf.

Now, let's examine a specific function, the function f shown in **Figure 4.2**. Notice that the processing rule is given by the expression $x^2 - 3x - 4$, which describes the algebraic process by which the input variable x is transformed into output. Processing rules do not have to be algebraic in nature to have a function, but algebraic rules are those most commonly found in elementary calculus. When we write $f(x) = x^2 - 3x - 4$, we are stating that the output variable $f(x)$ is obtained by first inserting the given x into $x^2 - 3x - 4$ and then calculating the result.

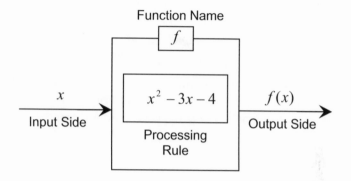

Function Name

f

x

Input Side

$x^2 - 3x - 4$

Processing
Rule

$f(x)$

Output Side

Figure 4.2: Function Process for $f(x) = x^2 - 3x - 4$

Ex 4.1.1: Calculating outputs when inputs are specific numbers.

$$f(0) = (0)^2 - 3(0) - 4 \Rightarrow f(0) = -4$$
$$f(1) = (1)^2 - 3(1) - 4 \Rightarrow f(1) = -6$$
$$f(-7) = (-7)^2 - 3(-7) - 4 \Rightarrow f(-7) = 66$$

Ex 4.1.2: Calculating outputs when inputs are variables or combinations of variables (see *pronoun number* or *'pronumber'* as explained in **Appendix A**).

$$f(a) = a^2 - 3a - 4 \text{, no further simplification possible}$$
$$f(a+h) = (a+h)^2 - 3(a+h) - 4 = a^2 + 2ah + h^2 - 3a - 3h - 4$$

Ex 4.1.3: Algebraic simplification where functional notation is part of the algebraic expression.

$$\frac{f(a+h) - f(a)}{h} = \frac{(a+h)^2 - 3(a+h) - 4 - \{a^2 - 3a - 4\}}{h} \Rightarrow$$
$$\frac{f(a+h) - f(a)}{h} = \frac{2ah + h^2 - 3h}{h} = 2a + h - 3$$

Now we consider Df and Rf for $f(x) = x^2 - 3x - 4$. Since we can create an output for any real number utilized as input, $Df = (-\infty, \infty)$, the interval of all real numbers. More extensive analysis is required to ascertain Rf. From algebra, $f(x)$ is recognized as a quadratic function having a *low point* on the vertical axis of symmetry as shown in **Figure 4.3**. The axis of symmetry is positioned midway between the two roots -1 and 4 where the associated x value is $\frac{3}{2}$. Using $\frac{3}{2}$ as input to f, we have that $f(\frac{3}{2}) = -\frac{25}{4}$, which is the smallest output value possible. Hence, $Rf = [-\frac{25}{4}, \infty)$ reflecting the fact that functional outputs for f (whose values are completely determined by input values) can grow increasingly large without bound as x moves steadily away from the origin in either direction.

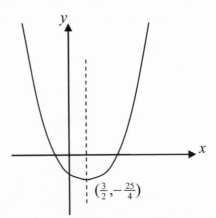

Figure 4.3: Graph of $f(x) = x^2 - 3x - 4$

Using the notation $f(x) = x^2 - 3x - 4$ is somewhat cumbersome at times. How can we shorten this notation? As previously stated, when it is understood that a function is in place and operating, we can write $y = x^2 - 3x - 4$ to represent the totality of the function process. The only change is that y replaces $f(x)$ as the name for the output variable. Other names for the output variable such as u, v, w are also used.

The simplicity that the x, y notation brings to the study of calculus will become evident as the book progresses. An immediate advantage is that one can readily plot input-output pairs associated with the function as (x, y) coordinates in a rectangular coordinate system. This plotting process is called graphing, and the finished product is called a graph as shown in **Figure 4.3**.

Ex 4.1.4: Find Dg and Rg for $g(t) = \dfrac{\sqrt{t}}{t-1}$. In this example, t is the input variable, and $g(t)$ is the output variable. Notice that the processing rule will not process negative t values or a t value of 1. This implies that the function g simply won't function for these input values. Hence, we must restrict Dg to $[0,1) \cup (1,\infty)$ so we don't get into *input trouble*. The domain Dg is called the *natural domain* for g since g will produce an output for any value chosen from $[0,1) \cup (1,\infty)$. The reader can verify that $Rg = (-\infty, \infty)$ by letting a be any proposed output value in the interval $(-\infty, \infty)$, setting $\dfrac{\sqrt{t}}{t-1} = a$, and solving for the input value t that makes it so. The t will always be found—guaranteed.

We now turn to the algebra of functions. Let f and g be any two functions. We can add, subtract, multiply, and divide these functions to create a new function F. This is done by simply adding, subtracting, multiplying, or dividing the associated processing rules as the following example shows. Once F is created, DF will have to be re-examined, especially in the case of division. Since RF is associated with the dependent variable, it usually doesn't need such detailed examination.

Ex 4.1.5: Let $f(x) = x-1$ & $g(x) = x^2$, $Df = Dg = (-\infty, \infty)$
To create:
$F = g + f$, set $F(x) = g(x) + f(x) = x^2 + x - 1$
$F = g - f$, set $F(x) = g(x) - f(x) = x^2 - x + 1$

$F = g \cdot f$, set $F(x) = g(x) \cdot f(x) = x^3 - x^2$

$F = \dfrac{g}{f}$, set $F(x) = \dfrac{g(x)}{f(x)} = \dfrac{x^2}{x-1}$

Notice that $DF = Df = Dg$ in the case of addition, subtraction, and multiplication. However, for division, $DF = (-\infty, 1) \cup (1, \infty)$.

Our last topic in this section is function composition, *which is input/output processing via a series of stages.* Let f and g be two functions. Consider the expression $f(g(x))$. Peeling back the functional notation to the core (so to speak) exposes an input variable x, input associated with the function g. The output variable for g is $g(x)$. But what dual role is $g(x)$ serving? Look at its position within the expression $f(g(x))$. Notice that $g(x)$ also serves as the input variable to the function f. Hence, outputs from g are the inputs to f; and the expression $f(g(x))$ is the final output variable from this two-stage process. The flow diagram below depicts the stages by which x is processed into $f(g(x))$.

$$\overset{1}{\mapsto} : x \underset{g}{\longrightarrow} g(x) \overset{2}{\mapsto} : g(x) \underset{f}{\longrightarrow} f(g(x))$$

$f(g(x))$ may also be written $f \circ g(x)$ where the first function encountered in normal reading of the symbol (in this case f) is the last stage of the process. Nothing prohibits the reversing of the two stages above. We can just as easily study $g(f(x))$ or $g \circ f(x)$ by constructing an appropriate flow diagram.

$$\overset{1}{\mapsto} : x \underset{f}{\longrightarrow} f(x) \overset{2}{\mapsto} : f(x) \underset{g}{\longrightarrow} g(f(x))$$

Be aware that the final output variable $f(g(x))$ is rarely the same as $g(f(x))$. But we shouldn't expect this. A light-hearted illustration clarifies: dressing the foot in the morning is a two-stage process.

First, one puts on a sock; and, secondly, one puts on a shoe—normal and accepted practice. The result is a product that results in comfortable walking. But, what happens when the process is reversed? One wears out socks at an extraordinary rate, and one's feet become very rank due to direct exposure to leather!

Ex 4.1.6: Form the two function compositions $f(g(x))$ and $g(f(x))$ if $f(x) = x^2 + 1$ and $g(x) = x - 1$.

$$f(g(x)) = f(x-1) = (x-1)^2 + 1 = x^2 - 2x = f \circ g(x)$$
$$g(f(x)) = g(x^2+1) = (x^2+1) - 1 = x^2 = g \circ f(x)$$

Notice $f \circ g(x) \neq g \circ f(x)$ (remember the shoe and sock).

Functional compositions can consist of more than two stages as the next example illustrates.

Ex 4.1.7: Let $f(x) = \dfrac{x+2}{3x-1}$, $g(x) = x^2$ and $h(x) = \sqrt{x}$.

Part 1: Construct a flow diagram for the composed (or staged) function H where $H = h \circ f \circ g$. Notice how the notation for the input variable is sometimes suppressed when speaking of functions in global terms.

$$\overset{1}{\mapsto} : x \xrightarrow[g]{} g(x) \overset{2}{\mapsto} : g(x) \xrightarrow[f]{} f(g(x))$$
$$\overset{3}{\mapsto} : f(g(x)) \xrightarrow[h]{} h(f(g(x)))$$

Part 2: Find the processing rule for $H(x) = h(f(g(x)))$.

Reader challenge: Find the associated DH and compare to Dh, Df and Dg.

$$H(x) = h(f(g(x))) = h(f(x^2)) = h\left(\frac{x^2+2}{3x^2-1}\right) = \sqrt{\frac{x^2+2}{3x^2-1}}$$

Part 3: Find the processing rule for $F(x) = f(g(h(x)))$.

Reader challenge: compare DF to DH in Part 2.

$$F(x) = f(g(h(x))) = f(g(\sqrt{x})) = f((\sqrt{x})^2) =$$

$$f(|x|) = \frac{|x| + 2}{3|x| - 1}$$

$$\int_a^b \ddot{\cup} \, dx$$

Section Exercise

Let $f(x) = x^3 - 4x$ and $g(x) = \sqrt{x}$.

A. Find processing rules for $f + g$, $f - g$, gf, $\dfrac{f}{g}$ and $\dfrac{g}{f}$.

B. Find the natural domain for each of the functions in A.

C. Find $f \circ g$ and $g \circ f$; find the natural domain for each.

D. Complete the table below for $f(x)$

Input Value	Output Value
2	
6	
0	
7	
3	
a	
$a + h$	

E. Simplify the expression $\dfrac{f(a+h) - f(a)}{h}$.

4.2) Inverse Functions: Output to Input

Suppose two functions f and g have the following composition behavior $g(f(x)) = x$. What does this mean? To help answer this question, first construct a flow diagram for the two-stage process.

$$\mapsto : x \xrightarrow{\ f\ } f(x) \mapsto : f(x) \xrightarrow{\ g\ } g(f(x)) = x$$

From the diagram, we see that x is not only the initial input but the final output. Let's follow the action stage by stage.

$\overset{1}{\mapsto}$: *Stage 1*

$x \xrightarrow{\ f\ } f(x)$: The input x is transformed into output $f(x)$ by f .

$\overset{2}{\mapsto}$: *Stage 2*

$f(x) \xrightarrow{\ g\ } g(f(x)) = x$: The output $f(x)$ now becomes input for g .The function g transforms $f(x)$ back into the original input x (as shown by the concise notation $g(f(x)) = x$).

Definition: Let f be any function and suppose there is a function g with the property that $g(f(x)) = x$. We then call g the *inverse function* of f and give it the new notation f^{-1}. Hence $g = f^{-1}$ by definition, and $f^{-1}(f(x)) = x$. The function f^{-1} is called the inverse function because it reverses (or undoes) the processing action of f by transforming outputs back to original inputs.

Ex 4.2.1: Let $g(x) = \dfrac{3x+1}{x-2}$ and $f(x) = \dfrac{2x+1}{x-3}$.

A) Show $g(f(4)) = 4$.

$$g(f(4)) = g\left(\frac{2(4)+1}{4-3} \right) = g(9) = \frac{3(9)+1}{9-2} = 4$$

B) Show $g = f^{-1}$.

$$g(f(x)) = g(\tfrac{2x+1}{x-3}) = \frac{3(\tfrac{2x+1}{x-3})+1}{(\tfrac{2x+1}{x-3})-2} = \frac{7x}{7} = x \Rightarrow g = f^{-1}$$

C) Show $f(f^{-1}(x)) = x$

$$f(f^{-1}(x)) = f(\tfrac{3x+1}{x-2}) = \frac{2(\tfrac{3x+1}{x-2})+1}{(\tfrac{3x+1}{x-2})-3} = \frac{7x}{7} = x$$

Note: B) and C) together imply $f^{-1}(f(x)) = f(f^{-1}(x)) = x$.

What conditions need to be in place in order for a function f to have an inverse? We will examine the function $f(x) = x^2$ in order to answer this question. Let $x = 2$. If f^{-1} exists, then we must have that $f^{-1}(f(2)) = f^{-1}(4) = 2$. But, we could also have that $f^{-1}(4) = -2$. The supposed function f^{-1} can transform the output 4 back to two distinct inputs, 2 and -2 (violating the very definition of what is meant by the word function). So, how do we guarantee that f has an inverse? Answer: we must restrict f to a domain where it is *one-to-one*.

Definition: A *one-to-one* function is a function where every unique input leads to a unique output. *Note: you are encouraged to compare and contrast this definition with the general definition of a function where any given input leads to a unique output.*

A one-to-one function f (sometimes denoted by $f_{1 \to 1}$) allows for precise traceability of each and every output back to a unique input. As a result, a true action-reversing function f^{-1} can be formulated, undoing the forward action of f.

Ex 4.2.2: By restricting the domain of $f(x) = x^2$ to the half interval $[0,-\infty)$, we create a one-to-one function. The domain-restricted function f now has an inverse given by $f^{-1}(x) = \sqrt{x}$.

Once a function f has been established as one-to-one on a suitable domain, we are guaranteed f^{-1} exists and has the property $f^{-1}(f(x)) = x$. An immediate consequence is the property $f^{-1}(f(x)) = f(f^{-1}(x)) = x$, a composition property unique to functions in a f, f^{-1} relationship. We will use this last property to actually find a processing rule for f^{-1} given a known processing rule for $f_{1\to1}$. What follows is a four step algebraic procedure for finding f^{-1}.

$\overset{1}{\mapsto}$: Start with $f(f^{-1}(x)) = x$, the process equality that must be in place for an inverse function to exist.

$\overset{2}{\mapsto}$: Replace $f^{-1}(x)$ with y to form the equality $f(y) = x$, the second use in this book of simplified output notation.

$\overset{3}{\mapsto}$: Solve for y in terms of x. The resulting y is $f^{-1}(x)$.

$\overset{4}{\mapsto}$: Verify by the property $f^{-1}(f(x)) = f(f^{-1}(x)) = x$.

The next two examples illustrate the above procedure.

Ex 4.2.3: Find $f^{-1}(x)$ for $f(x) = \dfrac{2x-3}{4x+7}$.

$$\overset{1}{\mapsto}: f(f^{-1}(x)) = \frac{2f^{-1}(x)-3}{4f^{-1}(x)-7} = x$$

$$\overset{2}{\mapsto}: \frac{2y-3}{4y-7} = x$$

$$\overset{3}{\mapsto}: 2y-3 = (4y+7)x \Rightarrow 2y-3 = 4yx+7x$$
$$\Rightarrow 2y-4yx = 7x+3 \Rightarrow y(2-4x) = 7x+3$$
$$\Rightarrow y = \frac{7x+3}{2-4x} = f^{-1}(x)$$

$\overset{4}{\mapsto}$ (Step4): Left as a reader challenge

Ex 4.2.4: Find $f^{-1}(x)$ for $f(x) = x^3 + 2$.

$\overset{1}{\mapsto}: f(f^{-1}(x)) = (f^{-1}(x))^3 + 2 = x$

$\overset{2}{\mapsto}: (y)^3 + 2 = x$

$\overset{3}{\mapsto}: (y)^3 + 2 = x \Rightarrow (y)^3 = x - 2 \Rightarrow y = \sqrt[3]{x-2}$

$\overset{4}{\mapsto}: f^{-1}(f(x)) = \sqrt[3]{(x^3 + 2)} - 2 = \sqrt[3]{x^3} = x$

$\overset{4}{\mapsto}: f(f^{-1}(x)) = \left(\sqrt[3]{(x-2)}\right)^3 + 2 = (x-2) + 2 = x$

$$\int_a^b \overset{..}{\cup} dx$$

Section Exercise

Consider the following three functions $f(x) = \dfrac{3x+5}{11}$,

$g(x) = \dfrac{2-x}{8-x}$ and $h(x) = \dfrac{x^2-7}{5}$.

A) What is the domain for each function?

B) Find a suitable restriction for the domain of each function so that the function is one-to-one.

C) Find f^{-1}, g^{-1} and h^{-1}.

4.3) Arrows, Targets, and Limits

When arithmetic is expanded to include variables and associated applications, arithmetic becomes algebra. Likewise, when algebra is expanded to include limits and associated applications, algebra becomes calculus. Thus, we can say, *the limit concept distinguishes calculus from algebra*. We can also say—as evidenced by the continuing sales success of Silvanus Thompson's 100-year-old self-help book, <u>Calculus Made Easy</u>— *that the limit concept is the main hurdle preventing a successful study of calculus*. An interesting fact is that excessive use of limits in presenting the subject of calculus is a relatively new thing (post 1950). Calculus can, in part, be explained and developed using an older—yet still fundamental—concept, that of differentials (e.g. as done by Thompson). In this book, the differential concept is the primary concept by which the subject of calculus is developed. Limits will only be used when absolutely necessary, but limits will still be used. So, to start our discussion of limits, we are going to borrow some ideas from the modern military, ideas that Thompson never had access to.

What is a limit? Simply put, a limit is a numerical target that has been acquired and locked. Consider the expression $x \rightarrow 7$ where x is an independent variable. The arrow $(\rightarrow)$ points to a target on the right, in this case the number 7. The variable x on the left is targeting 7 in a *modern smart-weapon sense*. This means x is moving, is moving towards target, is closing range, and is programmed to eventually merge with the target. Notice that x is a true variable: in that it has been launched and set in motion towards a target, a target that cannot escape from its sights. Now, our independent variable usually finds itself embedded inside an algebraic (or transcendental) expression of some sort, which is being used as a processing rule for a function. Consider the expression $2x + 3$ where the independent variable x is about to be sent on the mission $x \rightarrow -5$. Does the entire expression $2x + 3$ in turn target a numerical value as $x \rightarrow -5$? A way to phrase this question using a new type of mathematical notation might be

$$t \arg et (2x + 3) = ?$$
$$x \rightarrow -5$$

Interpreting the notation, we are asking if the output stream from the expression $2x+3$ targets a numerical value in the modern smart-weapon sense as x targets -5? Mathematical judgment says yes; the output stream targets -7. Hence, we complete our new notation as follows:

$$\underset{x \to -5}{t\arg et}(2x+3) = -7 .$$

This all sounds great except for one little problem: the word *target* is nowhere to be found in calculus texts. The traditional replacement (weighing in with 300 years of history) is the word *limit*, which leads to the following definition:

Definition: A *limit* is a target in the modern *smart-weapon* sense.

Correspondingly, our new *target* notation can be appropriately altered by writing $\underset{x \to -5}{\lim it}(2x+3) = -7$, and further shortened to $\underset{x \to -5}{\lim}(2x+3) = -7$. Let's investigate three limits using our new notation per the following example.

Ex 4.3.1: Suppose $f(x) = \dfrac{x^2 - 4}{x - 2}$. Evaluate the three limits:

A) $\underset{x \to 3}{\lim}(f(x))$, B) $\underset{x \to 2}{\lim}(f(x))$ and C) $\underset{x \to a}{\lim}(f(x))$.

A) $\underset{x \to 3}{\lim}(f(x)) = \underset{x \to 3}{\lim}(\frac{x^2-4}{x-2}) = 5$. Here, we just *slipped* the input target 3 into the expression $\dfrac{x^2 - 4}{x - 2}$ to obtain the output target 5.

Note: this is easily done by mathematical judgment, the mathematician's counterpart to engineering judgment.

B) Our judgment fails for $\underset{x \to 2}{\lim}(f(x))$ since a simple *slipping in* of 2 for x creates a division by zero. Here, we will need to return to the basic definition of limit or target. Recall that a target value is the value acquired, locked, and programmed *to be* merged with.

$$\lim_{x \to 2}(f(x)) = \lim_{x \to 2}\left(\frac{x^2 - 4}{x - 2}\right) =$$

Hence:

$$\lim_{x \to 2}\left(\frac{[x+2][x-2]}{x-2}\right) = \lim_{x \to 2}(x+2) = 4$$

In the above *equality stream*, the input x has acquired 2 and has locked on its target as denoted by $x \to 2$. Turning our attention to the associated output expression $\dfrac{x^2 - 4}{x - 2}$, we find that it algebraically reduces to $x + 2$ for all x traversed in the locking sequence $x \to$. The reduced expression $x + 2$ allows us to readily ascertain what the associated output stream has acquired as a target, which is 4. Targets are just that—targets! The mission is not always completed just because the output stream has acquired and locked a target. In some cases, the mission aborts even though the output stream is on a glide path to that eventual merging. In the case above, the mission has been aborted by division by zero—right at $x = 2$. However, it makes no difference; by definition, 4 is still the target.

C) In $x \to a$, the independent variable x has targeted the pronoun number a (perfectly acceptable under our definition since the target is a numerical value). As a result, the output stream targets an algebraic expression with

$$\lim_{x \to a}(f(x)) = \lim_{x \to a}\left(\frac{x^2 - 4}{x - 2}\right) = \frac{a^2 - 4}{a - 2}; a \neq 2.$$

Ex 4.3.2: Let $f(x) = x^3 - 4x$. Evaluate $\lim_{h \to 0}\left[\dfrac{f(a+h) - f(a)}{h}\right]$.

Two things are readily apparent. Just slipping in $h = 0$ creates the indeterminate expression $\frac{0}{0}$; so, just slipping in won't do. Also, the input target being a pronoun number probably means the output target will be an algebraic expression.

First, we simplify the expression $\dfrac{f(a+h)-f(a)}{h}$ before the limit is investigated, which is done to remove the perceptual problem at $h = 0$.

$$\frac{f(a+h)-f(a)}{h} = \frac{(a+h)^3 - 4(a+h)-[a^3-4a]}{h} =$$

$$\frac{a^3+3a^2h+3ah^2+h^3-4a-4h-a^3+4a}{h} = 3a^2+3ah+h^2-4$$

Notice that the perceptual problem has now been removed allowing one to complete process of finding the limit

$$\lim_{h\to 0}\left[\frac{f(a+h)-f(a)}{h}\right] = \lim_{h\to 0}[3a^2+3ah+h^2-4]=3a^2-4.$$

Independent variables within limits or targets do not need to be restricted to finite values. When we write $x \to \infty$, we mean that x is continuously inflating its value with no upper bound. A mathematical subtlety is that the infinity symbol ∞ really represents a process, the process of steadily increasing without bound with no backtracking. Hence, the expression $x \to \infty$ is redundant. There is no actual number at ∞: just a steady succession of process markers that mark the continuing unbounded growth of x.

Ex 4.3.3: Let $f(x) = 2+\dfrac{1}{x}$. Then $\lim\limits_{x\to\infty}\left[2+\dfrac{1}{x}\right] = 2$.

Ex 4.3.4: For $f(x) = \dfrac{2x+1}{1-3x}$, find $\lim\limits_{x\to\infty}[f(x)]$.

$$\lim_{x\to\infty}[f(x)] = \lim_{x\to\infty}\left[\frac{2x+1}{1-3x}\right] = \lim_{x\to\infty}\left[\frac{2+\frac{1}{x}}{\frac{1}{x}-3}\right] = -\frac{2}{3}.$$

Before moving on to our final example, a financial application, we'll plug a few notational holes. The notation $x \to -\infty$ simply means the process of decreasing without bound (i.e. the bottom drops out). Also, if we have that $y = f(x)$, the use of y again simplifies notation:

$$x \to a \Rightarrow y \to L \text{ and } \lim_{x \to a}[f(x)] = L \text{ mean the same thing.}$$

Ex 4.3.5: Consider the compound interest formula $A = P\left(1 + \frac{r}{n}\right)^{nt}$. Investigate $\lim_{n \to \infty}[p(1 + \frac{r}{n})^{nt}]$ given a fixed annual interest rate r and total time period t in years. The independent variable n is the number of compounding periods per year.

To solve this problem, we first move the limit process inside so the process can join the associated independent variable n :

$$\lim_{n \to \infty}[p(1 + \tfrac{r}{n})^{nt}] = p\{\lim_{n \to \infty}[(1 + \tfrac{r}{n})^{n}]\}^{t}.$$

Here, we have a classic battle of opposing forces. Letting an exponent go unbounded means that the quantity $a > 1$ to which the exponent is applied also goes unbounded $\lim_{n \to \infty}[a^{n}] = \infty$. However, when $a = 1$, the story is different with $\lim_{n \to \infty}[1^{n}] = 1$. In the expression above, the exponent grows without bound and the base gets ever closer to 1. What is the combined effect? To answer, first define $m = \frac{n}{r} \Rightarrow n = rm$. From this, we can establish the *towing relationship* $n \to \infty \Leftrightarrow m \to \infty$. Substituting, one obtains

$$\lim_{n \to \infty}[p(1 + \tfrac{r}{n})^{nt}] = p\{\lim_{n \to \infty}[(1 + \tfrac{r}{n})^{n}]\}^{t} = p\{\lim_{m \to \infty}[(1 + \tfrac{1}{m})^{m}]\}^{rt}.$$

Now let's examine $\lim_{m \to \infty}[(1 + \frac{1}{m})^{m}]$. We will evaluate it the easy way, via a modern scientific calculator.

m value	$(1+\frac{1}{m})^m$
1	2
10	2.5937
100	2.7048
1000	2.7169
10000	2.7181
100000	2.7183
1000000	2.7183

We stopped the evaluations at $m = 1,000,000$. Some might say that we are just getting started on the road to ∞. But we quit. Why? Look at the output stream; each time m is increased by a factor of 10, one more digit to the left of the decimal point is stabilized. Let's call it a day for $m = 1,000,000$ since four digits to the left of the decimal point would be quite good enough for most ordinary applications. If one needs a few more digits, one can always compute a few more digits. The actually result is the famous number $e = 2.7183...$, and our final limit becomes:

$$A = p\{\lim_{m\to\infty}[(1+\tfrac{1}{m})^m]\}^{rt} = p\{e\}^{rt}.$$

The expression $A = pe^{rt}$ is called the *continuous interest formula* (much more on this in Chapter 8). For a fixed annual interest rate r and initial deposit P, the formula gives the account balance A at the end of t years under the condition of continuously adding to the current balance the interest earned in a twinkling of an eye. The continuous interest formula represents in itself an upper limit for the growth of an account balance given a fixed annual interest rate. This idea will be explored further in the section exercises.

This ends our initial discussion of limits. We will visit limits again (though sparingly) throughout the book in order to help formulate some of the major results distinguishing calculus from algebra—again, only to be done on an as-needed basis.

To summarize:

> **Limits are foundational to calculus and always will be so.**
> **Limits lead to results unobtainable by algebra alone.**

$$\int_a^b \cup \, dx$$

Section Exercises

1. Evaluate: A) $\lim_{h \to 0}\left[\dfrac{(3+h)^2 - 9}{h}\right]$,   B) $\lim_{x \to \infty}\left[\dfrac{x^2 + 3x + 1}{5x^2 + 10}\right]$,

         C) $\lim_{x \to 5}[4x^2 - 7]$ and   D) $\lim_{x \to \infty}\left[2 + \dfrac{1000}{x}\right]$

2. Evaluate $\lim_{n \to \infty}\left(1 + \dfrac{k}{n}\right)^n$, $k > 1$. *This is more difficult!*

3. For $f(x) = 3x^2 - 7$, evaluate $\lim_{h \to 0}\left[\dfrac{f(a+h) - f(a)}{h}\right]$.

4. Compare the final amount from $A = pe^{rt}$ to the final amount from $A = p(1 + \frac{r}{n})^{nt}$ for $r = 7\%$, $t = 10$, and $p = \$2500.00$. In $A = p(1 + \frac{r}{n})^{nt}$, use the following compounding periods: $n = 2, 4, 6$ and 12 times annually.

4.4) Continuous Functions

Time in many ways is a mysterious concept, even though our modern lives are essentially governed by intervals of time. Let's examine one of these intervals, say the interval from 08:00 to 17:00 (military time) defining a typical workday. We could write this interval as $[8,17]$ and graph it on an axis labeled t.

Borrowing the stick person from Chapter 3, let's walk through our daily routine during these wide-awake working hours.

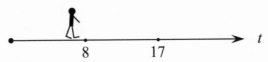

Several things are apparent: Our walk moves forward in the direction of increasing time, our walk does not stop, and our walk passes through all intermediate times when going from 8 to 17. The last statement says we cannot wave a magic wand at 9:55 and it suddenly become noon, skipping the dreaded 10:00 meeting with the boss. The nature of real life is that, in order to get from time A to time B, we must pass through all times in between. This "passing through" characteristic allows us to say that time *continuously flows* from a present value to a future value. The x axis behaves in exactly the same way. In the diagram below, the only difference is that the axis has been relabeled.

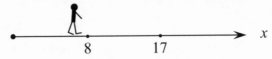

Our stick person starting a walk at 8 still must pass through all intermediate numbers in order to arrive at 17. By this simple illustration, it is easy to see that the entire x axis has the *continuous-flow* characteristic—just like time. The only real-life difference is that our figure's walking motion on the x axis can occur from left to right (like time) or from right to left (unlike time).

Now consider the function $h(t) = 3000 - 16.1t^2$ where time t in seconds is the input (independent) variable. The output or dependent variable $h(t)$ is the height of a free-falling object (neglecting air resistance) dropped from an altitude of 3000 feet. In the above mathematical model, $t = 0$ corresponds to the time of object release. The object does not fall indefinitely. Soon it will impact the earth at a future time T when $h(T) = 0$. Using the equality $h(T) = 0 \Rightarrow 3000 - 16.1T^2 = 0$, one can solve for T to obtain $T = 13.65$. Per the previous discussion on time, we can assume that time in the interval $[0, 13.65]$ flows continuously from $t = 0$ (release) to $t = 13.65$ (impact). *Likewise*, the output variable drops continuously from $h = 3000$ (release) to $h = 0$ (impact), skipping no intermediate altitudes on the way down. The function $h(t)$ is graphed in **Figure 4.4** where the *unbroken curve* signifies that all intermediate attitudes are traversed on the way to impact.

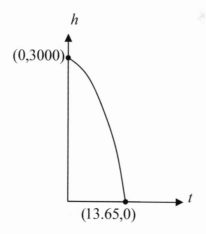

Figure 4.4: Graph of $h(t) = 3000 - 16.1t^2$

Any break or hole in the curve would mean that the object has mysteriously leaped around an intermediate time and altitude on the way to impact as shown in **Figure 4.5**—quite an impossibility.

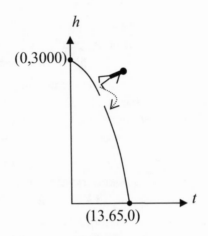

Figure 4.5: The Impossible Leap

Let's remove the real-world context. This is easily done by writing $f(x) = 3000 - 16.1x^2$ and treating $f(x)$ solely as an object of algebraic study. Expand the domain to the natural domain and graph $f(x)$. *Note: the variable t in $h(t)$ cannot be negative since t is representing the real-world phenomena of time. However, the variable x in $f(x)$ can be negative since it (employed as an object of algebraic study) is not being used to represent anything in the real world.*

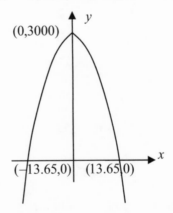

Figure 4.6: Graph of $f(x) = 3000 - 16.1x^2$

Figure 4.6 is the result. Notice that the enlarged graph is identical in part to the graph in **Figure 4.4**. The two inclusions (due to expansion of the natural domain to all real numbers) are 1) a left side and 2) negative y values. Since the entire x axis serves as the natural domain and has the *continuous-flow* characteristic, the output stream should also have the *continuous-flow* characteristic when transitioning from one value to another. Again, *continuous flow* is the mathematical characteristic that allows us to graph $f(x) = 3000 - 16.1x^2$ as an unbroken curve on its natural domain. The following is a working definition for a *continuous function f* .

<u>Definition</u>: let f be defined on a subinterval $[a,b]$ of the x axis. We say f is continuous on $[a,b]$ if the output values from f flow smoothly (no gaps or jumps) on a path from $f(a)$ to $f(b)$ as the input values flow from a to b . Note: this definition will allow the flow path from $f(a)$ to $f(b)$ to meander, but the flow path can not have a break or chasm.

The good news is that most functions used to model real-world phenomena are continuous on their respective domains of interest. Notable simple and practical counterexamples of *discontinuous functions* exist, especially in the world of business and finance (Section Exercises). Algebraic functions are always continuous except 1) in those regions of the x axis where even roots—square roots, etc.—of negative inputs are attempted or 2) those single x values where a division by 0 occurs.

Ex 4.4.1: Find the region on the x axis where the algebraic function $f(x) = \sqrt{\dfrac{x+10}{x+2}}$ is continuous. The "root exception" 1) leads to the rational inequality

$$\frac{x+10}{x+2} \geq 0 \Rightarrow x \in (-\infty, -10] \cup [-2, \infty) .$$

Exception 2), division by 0 , is also not allowed. Thus, the *region of continuity* reduces further to $(-\infty, -10] \cup (-2, \infty)$.

To close this section, we will merge the informal definition of continuity with the formal definition of limit:

Definition: a function f is said to be continuous at a point $x = a$ if $\lim\limits_{x \to a}(f(x)) = f(a)$. Via direct implication, three things must happen for f to be continuous at $x = a$: the limit as $x \to a$ exists in the modern smart-weapon sense, the output $f(a)$ is defined, and the actual value of the limit is $f(a)$. Interpreting, the definition states that the output stream $f(x)$ targets $f(a)$ as x targets a and actually merges with $f(a)$ as x merges with a. All of this is just another way of saying *continuous flow*.

Ex 4.4.2: Use the above definition to show that the function $f(x) = \dfrac{x^2 - 4}{x - 2}$ is discontinuous at $x = 2$. We have that $\lim\limits_{x \to 2}(f(x)) = 2$ by **Ex 4.3.1**. However, $f(2)$ does not exist, which leads to $\lim\limits_{x \to 2}(f(x)) \neq f(2)$. Hence, by definition $f(x)$ is not continuous at $x = 2$. Note: A graph of $f(x)$ would have a little hole at $x = 2$ since $x = 2$ is not part of the natural domain of f.

$$\int_a^b \cup\, dx$$

Section Exercises

1. Find the region of continuity for $f(x) = \sqrt{\dfrac{x^2 - 2x}{x - 1}}$.

2. A parking garage charges $\$3.00$ for the first half hour and an additional $\$1.00$ for every additional half hour or fractional part thereof—not to exceed $\$10.00$ for one 12 hour period. Graph the parking charges versus time for one 12 hour period. Is this a continuous function on the time interval $[0,12]$? Why or why not?

4.5) The True Meaning of Slope

The concept of slope is usually associated with a straight line. In this section, we will greatly expand the traditional concept of slope and extract its true meaning—*that of a change ratio*. But first, let's review slope in its traditional setting via the straight line.

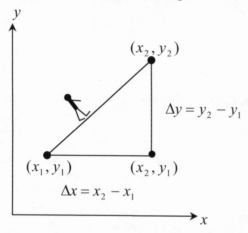

Figure 4.7: Line Segment and Slope

In **Figure 4.7**, our familiar stick person starts at the point (x_1, y_1) and *walks the line* to the point (x_2, y_2). In doing so, the figure experiences a change—denoted by the symbol Δ—in two dimensions. The change in the vertical dimension is given by $\Delta y = y_2 - y_1$. Likewise, the change in the horizontal dimension is given by $\Delta x = x_2 - x_1$. The slope m of the line is defined as the simple change ratio:

$$m = \frac{\Delta y}{\Delta x} = \frac{y_2 - y_1}{x_2 - x_1}.$$

Slope can be rendered in English as *so many y units of change per so many x units of change*. The two key words are *units* and *per* (the English rendering of the fraction bar) as now illustrated by the following two walks in dissimilar coordinate systems.

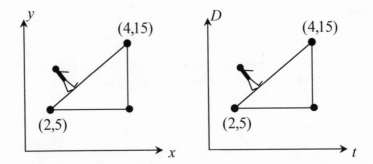

Figure 4.8: Similar Walks—Dissimilar Coordinates

In **Figure 4.8**, the stick person walks two visually-similar line segments in dissimilar coordinate systems. Granted, the numerical quantities may be identical in both cases, but the dimensions are different. In the leftmost coordinate system, the units assigned to both x and y are inches. In the rightmost coordinate system, the units assigned to D are miles and the units assigned to t are hours. Let's compute the slope m in both cases and interpret. In the leftmost walk:

$$m = \frac{\Delta y}{\Delta x} = \frac{(15-5)inches}{(4-2)inches} = \frac{10 inches}{2 inches} = 5\frac{inches}{inch}.$$

Here, the slope m is 10 inches of y change per 2 inches of x change. Dividing the 10 by 2 scales the y change to a single unit of x change, and the final slope can be rendered as 5 inches (of y change) per inch (of x change). When speaking of slope, the two phrases "of y change" and "of x change" are usually dropped. Dropping the phrase "of change" can lead to some potential confusion between y & Δy and x & Δx if we forget that we are dealing with a change ratio. In the rightmost walk:

$$m = \frac{\Delta D}{\Delta t} = \frac{(15-5)miles}{(4-2)hours} = \frac{10 miles}{2 hours} = 5\frac{miles}{hour}.$$

In this case, the slope m is 10 miles of y change per 2 hours of x change. Dividing the 10 by 2 again scales the y change to a single unit of x change, and the final slope can be rendered as 5 miles per hour. You should recognize this as the familiar expression for average velocity over a time interval.

Slope interpretation is always that of unit(s) of change per unit of change. Suppose the rightmost horizontal axis in **Figure 4.8** represents weeks (w), and the vertical axis, the number of houses (h) built by a major construction company. Then, the slope m would be:

$$m = \frac{\Delta h}{\Delta w} = \frac{(15-5)houses}{(4-2)week} = \frac{10houses}{2weeks} = 5\frac{houses}{week}.$$

Here, slope represents an average construction rate or average *construction velocity*. In formulating slopes or change ratios, the denominator is usually the total change in the assumed independent variable, represented by the horizontal axis. Likewise, the numerator is usually the corresponding total change in the assumed dependent variable, represented by the vertical axis. An actual cause-and-effect relationship between the independent and dependent variables strongly suggests that a function exists.

A straight line has the unique property that the slope or change ratio remains constant no matter where we are on the line.

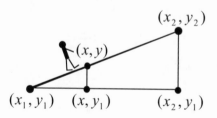

This *constant-slope* property can be used to develop the well-known point-slope equation of a straight line. Let (x, y) be any arbitrary point on the line.

Then by similar triangles, we have that

$$m = \frac{\Delta y}{\Delta x} = \frac{y - y_1}{x - x_1} = \frac{y_2 - y_1}{x_2 - x_1} \Rightarrow$$

$$y - y_1 = \left[\frac{y_2 - y_1}{x_2 - x_1} \right] (x - x_1) \Rightarrow$$

$$y - y_1 = m(x - x_1) \therefore$$

The equation $y - y_1 = m(x - x_1)$ can be easily simplified further to the slope-intercept form $y = mx + b$, where b is the y intercept and $-b/m$ is the x intercept. Notice that the slope-intercept form $y = mx + b$ describes a functional relationship between the input variable x and the output variable y.

Ex 4.5.1: Write the equation of a straight line passing through the two points $(-2,5)$ and $(3,-9)$. In actuality, it makes no difference which point is labeled (x_1, y_1) and which point is labeled (x_2, y_2). But, this book will use the following convention: (x_1, y_1) will be the point with the smallest x coordinate. Therefore, we have that $(x_1, y_1) = (-2,5)$, and $(x_2, y_2) = (3,-9)$ by default. The slope m (*units of y per unit of x*) is given by

$$m = \frac{[-9 - (5)]}{[3 - (-2)]} = \frac{-14}{5} = -\frac{14}{5}.$$

Once m is obtained, the equation of the line readily follows $y - 5 = -\frac{14}{5}(x + 2) \Rightarrow y = \frac{-14}{5}x - \frac{3}{5}$. Reducing to the slope-intercept *and functional* form $y = -\frac{14}{5}x - \frac{3}{5}$ allows for quick determination of both the x and y intercepts: $(-\frac{3}{14}, 0)$ and $(0, -\frac{3}{5})$. Sometimes, we say a line has three essential parameters: the slope, the x intercept, and the y intercept. To characterize a line means to find all three.

$$\int_a^b \overset{\cdot\cdot}{\cup} dx$$

Section Exercises

1. Write the equation of a line passing though the points $(6,-3)$ and $(-1,7)$ and characterize.

2. Write the equation of a line with slope $m = 3$ and passing through the point $(2,1)$. Characterize this line.

3. A traveler travels an Interstate highway for 5 hours starting at mile marker 200 and ending at mile marker 500. Plot these two points on a Distance versus time (t, D) coordinate system and calculate the slope for the line segment connecting the two points. Interpret the slope in terms of the traveler's velocity. In reality, what velocity does the slope represent? Assuming a linear (line-like) relationship, write the distance D as a function of time i.e. $D(t)$ for t in the domain $[0,5]$.

Riding the Beam

Captain Kirk, come beam me yonder
To a time of future wonder,
To the years where we could be
If Archimedes on his knee
Alone was left to dream and ponder.

But Rome then skewered from behind
And left us stranded with our mind
To contemplate what might have been
If earth just once could have a win
While on the rise with humankind

December 2001

Note: Archimedes, the inventor of a 'proto-calculus', died at the age of 75 in 212BC per Roman hands. His remarkable investigations were not to be continued again in earnest until the Renaissance.

4.6) Instantaneous Change Ratios

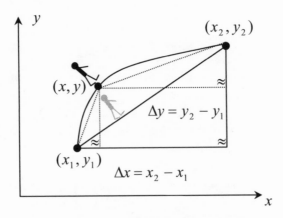

Figure 4.9: A Walk on the Curve

We are now ready to do an initial exploration of **The First Fundamental Problem of Calculus** as introduced in Chapter 3. Again, let our stick person walk the curve from (x_1, y_1) to (x_2, y_2) as shown in **Figure 4.9**. The overall Δy and Δx will be exactly the same as that experienced by the lighter-shaded stick person on the line segment below, where the slope remains constant. But, does the slope remain constant while walking the curve? The answer is no. Let (x, y) be any intermediate point on the curve. From the relative shapes of the two dotted triangles, we can easily see that average slope encountered from (x_1, y_1) to (x, y) is much greater than that encountered from (x, y) to (x_2, y_2). Also, by comparing these slopes to the three similar triangles marked with a $\approx$, we find that the *overall average slope* from (x_1, y_1) to (x_2, y_2) satisfies

$$\frac{y_2 - y}{x_2 - x} \leq \frac{y_2 - y_1}{x_2 - x_1} \leq \frac{y - y_1}{x - x_1}$$

for any point (x, y) on this particular walk.

Suppose the question is asked, what is the *exact slope* or *instantaneous change ratio* at the point (x, y) as depicted by the thickened triangle in **Figure 4.10**? For starters, at least three approximate answers are possible, represented by the three sides of the above inequality. Each answer depends on an arbitrary choice of two reference points and the slope formula for a straight line. Per visual inspection, none of these possibilities seem to match the exact slope experienced when our stick person is walking right on the point (x, y). Obviously, some fundamental improvement in methodology is needed.

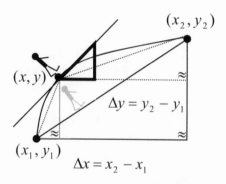

Figure 4.10: Failure to Match Exact Slope

In order to start developing our improved methodology, we let y be a function of x (i.e. $y = f(x)$). Accordingly, we re-label our walking-the-curve diagram as shown in **Figure 4.11**. The problem is to find the *exact slope* or *instantaneous change ratio* at the point $(a, f(a))$. Our precariously-perched stick person is definitely experiencing *exact slope* in terms of a greatly accelerated heart rate. But, can we compute it?

Continuing, place a second point $(a + h, f(a + h))$ on the curve where h is a true variable (moving quantity). When h is large, the point $(a + h, f(a + h))$ will be some distance from the point $(a, f(a))$ as shown in **Figure 4.11**.

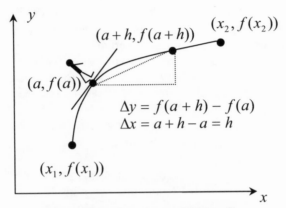

Figure 4.11: Conceptual Setup for Instantaneous or Exact Change Ratios

Collapsing h or letting h draw down to zero will in turn draw the point $(a+h, f(a+h))$ back towards—and eventually close to— the point $(a, f(a))$. In this scenario, $(a, f(a))$ is the fixed point and $(a+h, f(a+h))$ is the mobile point. No matter what value we choose for $h \neq 0$, we can calculate both Δy and Δx for the straight line segment connecting $(a, f(a))$ and $(a+h, f(a+h))$. We can also calculate the slope m for the same line segment, given by

$$m = \frac{\Delta y}{\Delta x} = \frac{f(a+h) - f(a)}{a+h-a} = \frac{f(a+h) - f(a)}{h}.$$

Recall that the slope m is in terms of vertical change over horizontal change: units of the dependent variable y per single unit of the independent variable x. The slope m is precise for the line segment, but only serves as an estimate for the slope of the curve at the point $(a, f(a))$, in that m is the average slope experienced when walking the curve from $(a, f(a))$ to $(a+h, f(a+h))$. So, how is the estimate improved?

We can use the limit process

$$\lim_{h \to 0}\left[\frac{\Delta y}{\Delta x}\right] = \lim_{h \to 0}\left[\frac{f(a+h) - f(a)}{h}\right]$$

to draw the mobile point $(a+h, f(a+h))$ ever closer and back into the fixed point $(a, f(a))$. As $h \to 0$, the steady stream of slopes $\dfrac{\Delta y}{\Delta x}$ produced should target and eventually merge with the exact slope as experienced by the stick figure at $(a, f(a))$. This *exact slope* or *instantaneous change ratio* at $(a, f(a))$ is denoted by $f'(a)$ (**Figure 4.12**). It is time for a very key definition.

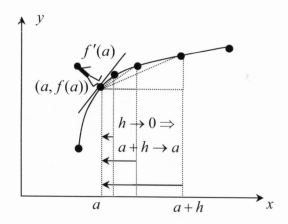

Figure 4.12: Better and Better Estimates for $f'(a)$

<u>Definition</u>: Let f be a function defined on an interval $[a,b]$ and let c be a point within the interval. The symbol $f'(c)$ is defined as the *exact slope* or *instantaneous change ratio* at $x = c$ provided this quantity exists. One way of obtaining the quantity $f'(c)$ is by investigating the limit

$$\lim_{h \to 0} \left[\frac{\Delta y}{\Delta x} \right] = \lim_{h \to 0} \left[\frac{f(c+h) - f(c)}{h} \right].$$

If the above limit exists, we will go ahead and define $f'(c)$ to be the value of this limit:

$$f'(c) = \lim_{h \to 0} \left[\frac{\Delta y}{\Delta x} \right] = \lim_{h \to 0} \left[\frac{f(c+h) - f(c)}{h} \right].$$

The quantity $f'(c)$ is known in words as the *first derivative* (or first derived function) of f at $x = c$. The process of obtaining $f'(c)$ is known as *differentiation*. One way—the most common way—of doing the process of differentiation is by using limits. Another way is by using differentials, to be discussed in chapter 5.

Ex 4.6.1: Using a limit process, differentiate the polynomial function $f(x) = x^2 - 3x - 4$ for any given x and interpret the result (refer back to **Ex 4.1.3**).

In this example x takes the place of c in the definition above and will remain fixed throughout the differentiation process—again, a process where h is the only true variable. But, once the process is completed, the product $f'(x)$ will be usable for any given x, providing *exact slope* as a function of x. By definition,

$$f'(x) = \lim_{h \to 0} \left[\frac{f(x+h) - f(x)}{h} \right] \Rightarrow$$

$$f'(x) = \lim_{h \to 0} \left[\frac{(x+h)^2 - 3(x+h) - 4 - \{x^2 - 3x - 4\}}{h} \right]$$

We will stop here for a moment and address an issue that frustrates those students who try to short-circuit the limit process by a direct substitution of $h = 0$ in the above or similar expression.

Notice that when you do this, you end up with

$$f'(x) = \left[\frac{(x+0)^2 - 3(x+0) - 4 - \{x^2 - 3x - 4\}}{0} \right] = \frac{0}{0},$$

an indeterminate expression. Why? Setting $h = 0$ immediately creates two identical points $(x, f(x))$ and $(x+0, f(x+0))$, from which no slope can be made since slopes require two points—in particular two points having distinct horizontal coordinates—for their formulation. So, continuing with the example,

$$f'(x) = \lim_{h \to 0} \left[\frac{2xh + h^2 - 3h}{h} \right] \Rightarrow f'(x) = \lim_{h \to 0} [2x + h - 3].$$

Notice that the problematic h now cancels into the numerator. This is quite acceptable since we are only interested in ascertaining the limit (target) as $h \to 0$, not whether the target is actually achieved when $h = 0$. Also, notice that the expression $2x + h - 3$ is always based on two points no matter how minuscule we make h, making it a valid expression for slope having the acceptable units of slope. Completing the differentiation process by use of limits, we have that

$$f'(x) = \lim_{h \to 0} [2x + h - 3] = 2x - 3.$$

Again $f'(x) = 2x - 3$, called the first derivative, is the product from the just-completed differentiation process. By definition, $f'(x)$ is the exact slope or instantaneous change ratio at an arbitrary point x in the domain of the function $f(x) = x^2 - 3x + 4$.

In **Ex 4.6.1**, $f'(x) = 2x - 3$ is a new function created from the associated *parent function* $f(x) = x^2 - 3x + 4$. We claim that $f'(x)$ is slope as a function of x. Let's check out this claim as to reasonableness by calculating three slopes in the domain of f (**Figure 4.13**).

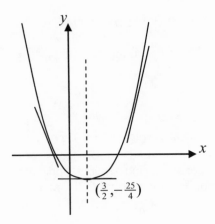

$\left(\frac{3}{2}, -\frac{25}{4}\right)$

Figure 4.13: Three Slopes for $f(x) = x^2 - 3x - 4$

The three points that we shall use are $x = -1.5$, $x = \frac{3}{2}$, and $x = 3$. Computing the exact slope $f'(x)$ for each point, we have the following: $f'(-1.5) = 2(-1.5) - 3 = -6$, $f'(\frac{3}{2}) = 0$, and $f'(4) = 5$. All three slopes seem to match the visual behavior of the graph in **Figure 4.13**. The slope is negative where it should be negative $(-\infty, \frac{3}{2})$ and positive where it should be positive $(\frac{3}{2}, \infty)$. In addition, the slope is 0 at $x = \frac{3}{2}$. This, the lowest point on the graph, is in the center of the valley and level by visual inspection.

One of the things that we can immediately do with our new-found slope information is to develop the equation of a *tangent line* at a point $x = c$, a line given by the formula

$$y - f(c) = f'(c)(x - c).$$

A tangent line is simply a line that meets the function f at the point $(c, f(c))$ and has exactly the same slope, $f'(c)$, as f at $x = c$. The three lines in **Figure 4.13** represent tangent lines.

The tangent-line formula is nothing more than the point-slope formula in **Section 4.5** modified for a *pre-determined* slope. At the point $x = -1.5$, the tangent-line equation is $y - f(-1.5) = f'(-1.5)(x + 1.5)$, which reduces after substitution to $y - 2.75 = -6(x + 1.5)$, which can be reduced further to the slope-intercept form $y = -6x - 6.25$. Lastly, the exact slope in this context is in terms of y units of change per single x unit of change. The next example shows how to adjust the interpretation of slope while maintaining the fundamental meaning.

Ex 4.6.2: Differentiate $D(t) = t^2 - 3t + 4$ and interpret by using the limit process where distance D is in feet, and time t is in seconds. This is exactly the same function that we had before except f is now D and x is now t, making the derivative $D'(t) = 2t - 3$. What is the interpretation? If we graph our function on a $D \& t$ coordinate system, the graph will have the same appearance as the graph in **Figure 4.13**, and $D'(t)$ will represent slope. But, $D'(t)$ is more than slope in that it has the units of feet per second, *which are the units of velocity*. Interpreting further, $D'(t)$ is not just an average velocity but *an instantaneous velocity* (like the velocity registered on your car's speedometer) for a given time t. As a final note, whenever a function f has a derivative f' that can be interpreted as a physical velocity, we can replace the f' notation with $\dot{f}$.

$$\int_a^b \ddot{\cup} \, dx$$

Section Exercise

1. Differentiate the function $g(x) = -7x^2 + 3x + 10$ and find the equation of the tangent line at 1) $x = -2$ and 2) the value of x where $f'(x) = 0$. Graph the function and the two tangent lines.

4.7) 'Wee' Little Numbers Known as Differentials

  Wee is a Scottish word that means *very small, tiny, diminutive, or minuscule*. *Wee* is a wonderful word that I have often used (in poetry) to describe something small when compared to something large. In the context of calculus, I will use it in similar fashion to help explain the concept of differential—also called an *infinitesimal*—which is the core concept providing the foundation for most of the mathematical techniques presented in this book.

  To have a differential, we must first have a variable, say x, y, z etc. Once we have a variable, say x, we can create a secondary quantity dx, which is called the differential of x. So, what exactly is this dx, read 'dee x'? The absolute value of the quantity dx by itself is a *very small, tiny, diminutive, or minuscule* numerical amount, symbolized by $0 < |dx| \ll 1$ where the left-skewed inequality can be interpreted: $|dx|$ resides in the open interval $(0,1)$ and is exceptionally close to the endpoint 0. It is the very small size of dx that makes it, by definition, a *wee* x. In addition, dx when compared to the original x is also *very small, tiny, diminutive, or minuscule*. How small? In algebraic terms, the following two inequalities hold:

$$0 < |x dx| \ll 1$$

$$0 < \left| \frac{dx}{x} \right| \ll 1.$$

The above two conditions state $|dx|$ is small enough to guarantee that the quantities $|x dx|$ and $\left| \frac{dx}{x} \right|$ also remain very small per the interpretation given above. Lastly, both inequalities imply $|dx| > 0$, which brings us to the following very important point:

> *Although very, very small,*
> *the quantity dx is never zero.*

One can also think of dx as the final h in a limit process $\lim\limits_{h \to 0}$ where the process abruptly stops just short of target—in effect, saving the rapidly vanishing h from disappearing into oblivion! Thinking of dx in this fashion makes it a prepackaged or frozen limit of sorts, an idea that will be explored again and again in this volume.

So, how are differentials used in calculus? They are primarily used to represent tiny increments of change. For example, suppose w is a variable; and dw, the associated differential. From these two quantities, $w + dw$ can be formulated, which represents the basic quantity w with just a wee bit, dw, added. In this example, the variable itself is changed from w to $w + dw$, and the tiny increment of change is the differential dw. Now suppose $y = f(x)$, and the independent variable is changed by addition of the differential dx. A natural question arises, what is the corresponding change in the dependent variable, denoted by the differential dy? Answer: it all depends on the processing rule associated with the function f as the following example augmented by **Figure 4.14** will show.

Ex 4.7.1: Calculate dy for $y = x^2$.

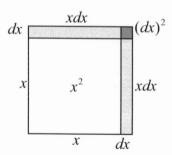

Figure 4.14: Differential Change Relationship for $y = x^2$

Solution: First, we create $x + dx$ by adding a differential increment dx to x.

This in turn *induces* a differential increment in y via the functional relationship $y = x^2$. This induced differential increment is the desired dy, which will be related to dx. We have that

$$y + dy = (x + dx)^2 \Rightarrow y + dy = x^2 + 2xdx + (dx)^2 \Rightarrow$$
$$y + dy - y = x^2 + 2xdx + (dx)^2 - x^2 \Rightarrow$$
$$dy = 2xdx + (dx)^2 \dots$$

The differential change relationship for the function $y = x^2$ can be diagramed using two nested squares as shown in **Figure 4.14**. The total area of the white square is $x^2 = y$ and the total area of the larger multi-partitioned square is $(x + dx)^2 = y + dy$. The induced differential dy, where $dy = 2xdx + (dx)^2$, is represented by the three grey-shaded areas: two lightly shaded with combined area $2xdx$ and one darkly shaded with area $(dx)^2$. Now, what can be said about $(dx)^2$ if $0 < |dx| << 1$? *Basically, $(dx)^2$ is so incredibly small that it must be totally negligible.* Hence, mathematicians will neglect the term $(dx)^2$ because it is of no consequence. Hence, the final answer for **Example 4.7.1** becomes $dy = 2xdx$: where dy, for a given x, is a linear function of dx in our differential or *infinitesimal* micro-world.

When dealing with differentials, we always assume that they are so incredibly small that second-order effects can be ignored. This leads to the following extension of our previous point concerning the differential dx (or $dy, dz, dw\dots$).

Although very, very small, dx is never zero.
But, it is still small enough to make $(dx)^2$ totally negligible.

The above is not only true for differentials associated with an independent variable (say x), but it is also true of dependent (induced) differentials related by $y = f(x)$.

For induced differentials our conceptual dx needs to be small enough so that the associated y and dy likewise satisfies the two fundamental conditions

$$0 < |ydy| << 1$$

$$0 < \left| \frac{dy}{y} \right| << 1.$$

Note: When I was in school, seasoned engineering and physics professors would say, dx needs to be small enough so that the associated differential dy behaves itself. Words like 'behave' make theoretical mathematicians cringe. But, over 200 years of engineering problem formulation via differentials has sent humankind both to the sea floor and to the moon. Let success speak for itself.

We will finish this section with two rules for differential arithmetic followed by three examples that illustrate the use of these rules.

1) Since $(dx)^2$ is negligible, $(dx)^n : n > 2$ is also totally negligible.

2) If $y = f(x)$, dy will also satisfy the two fundamental conditions as long as one makes dx small enough. Thus, $(dx + dy)^2$ is negligible. This last statement implies that each of the terms in $(dx)^2 + 2(dx)(dy) + (dy)^2$ is also negligible, in particular, the cross term $(dx)(dy)$.

Note: dependent differentials in a functional relationship given by $y = f(x)$ are usually written using the simplified y notation.

Ex 4.7.2: Find dy for $y = \sqrt{x}$.

$$y + dy = \sqrt{x + dx} \Rightarrow (y + dy)^2 = x + dx \Rightarrow$$

$$y^2 + 2ydy + (dy)^2 = x + dx \Rightarrow x + 2ydy = x + dx \Rightarrow$$

$$2ydy = dx \Rightarrow dy = \frac{dx}{2y} = \frac{dx}{2\sqrt{x}} \therefore$$

Ex 4.7.3: Find dy for $y = \dfrac{1}{x}$.

$$y + dy = \frac{1}{x + dx} \Rightarrow (y + dy)(x + dx) = 1 \Rightarrow$$

$$xy + xdy + ydx + dxdy = 1 \Rightarrow$$

$$1 + xdy + ydx = 1 \Rightarrow$$

$$xdy = -ydx \Rightarrow dy = \frac{-ydx}{x} = \frac{-dx}{x^2} \therefore$$

Ex 4.7.4: Find dy for $y = 4x^3$.

$$y + dy = 4(x + dx)^3 \Rightarrow$$

$$y + dy = 4(x^3 + 3x^2 dx + 3x(dx)^2 + (dx)^3) \Rightarrow$$

$$y + dy = 4x^3 + 12x^2 dx + 12x(dx)^2 + 4(dx)^3 \Rightarrow$$

$$4x^3 + dy = 4x^3 + 12x^2 dx + 12x(dx)^2 + 4(dx)^3 \Rightarrow$$

$$dy = 12x^2 dx + 12x(dx)^2 + 4(dx)^3 \Rightarrow$$

$$dy = 12x^2 dx \therefore$$

In each of the previous examples, the final result can be written as $dy = g(x)dx$ where $g(x)$ is a new function that has been derived from the original function $y = f(x)$. The fact that we can ignore all second order and higher differential quantities allows us to express dy as a simple linear multiple of dx for any given x.

$$\int_a^b \cup\!\!\!\cdot\!\cdot\, dx$$

Section Exercises

1. Find dy for $y = 13x + 5$ and interpret the $g(x)$ in $dy = g(x)dx$.

2. Find dy for A) $y = -7x^2 + 3x + 10$ and B) $y = \sqrt[3]{(2x - 5)}$.

4.8) A Fork in the Road

"Two roads diverged in a wood, and I—
I took the one less traveled by,
And that has made all the difference." Robert Frost

    I learned and grew up with calculus as it was traditionally taught after the collective shock of Sputnik I in October of 1957 (See Epilogue). Rigor was the battle cry of the day; and I, as a young man, took a lot of pride in mastering mathematical rigor. The whole body of calculus was built up via the route of definition, lemma, theorem, proof, example, and application—*if one was lucky enough to see an application*. Definitions were as impenetrable to the mathematically uninitiated as Egyptian Hieroglyphics would have been to me. An example of one such definition for a function f having a limit L at $x = a$ is

$$\lim_{x \to a} f(x) = L \Rightarrow \forall \varepsilon > 0, \exists \delta = \delta(\varepsilon)$$

$$s.t. f(x) \in N'(L, \varepsilon) \forall x \in N'(a, \delta)$$

Using modern jargon, you might call the above a "Please don't go there!" limit definition. We won't!

    In this book, we are going to travel backwards in time— pre 1957—to those years where the differential was the primary means by which the great concepts and powerful techniques of calculus came into play. And, we are really going back further than that.

For, our story starts with Sir Isaac Newton and Gottfried Leibniz. Each of these brilliant rivals independently invented the calculus towards the later part of the seventeenth century. However, neither of these men made use of intricate definitions or notations such as the limit definition shown on the previous page. Rigor was to come over a century later, primarily through the collective genius of Gauss, Cauchy, and Riemann. So what fundamental concept did Newton and Leibniz use to invent/create calculus? Differentials! Today, the notation $dx, dy, dz...$ etc. is still called Leibniz notation in honor of the man who first devised and used it.

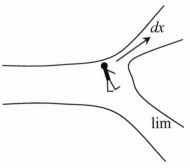

Figure 4.15: And that has made all the Difference...

As **Figure 4.15** indicates, this chapter ends at a pedagogical *fork in the road*: calculus taught by limits or calculus taught by differentials. The fork to be taken is *calculus taught by differentials*. A good thing is that the foundations, as set forth in this chapter, are essentially the same no matter which of the two roads we take. If we were teaching calculus by limits, much more rigor—especially in the sections on limits, continuity, and instantaneous change ratios—would have been needed in order to support subsequent theoretical developments. But, our road has been set, the road first traveled by Newton and Leibniz. A modern irony is that, even today, this road remains heavily traveled by professionals in engineering and physics who must formulate and solve the multitudinous *differential equations* (introduced in Chapter 6) arising in natural science. Often, they first learn the differential concept in the context of natural science—with very little support from a modern, limit-rich calculus. But in this book, the two great 17th century physicists, Newton and Leibniz, will again serve as our collective guides.

$$\int_a^b \overset{\cdot\cdot}{\cup} dx \quad \int_a^b \overset{\cdot\cdot}{\cup} dx$$

Chapter Review Exercise

Let $y = f(x) = x^4 - 2x^2$. Determine both dy and $f'(x)$. Find the equation of the tangent line at $x = -2$ and those points x where the tangent line is horizontal.

The Road Not Taken

Two roads diverged in a yellow wood,
And sorry I could not travel both
And be one traveler, long I stood
And looked down one as far I could
To where it bent in the undergrowth;

Then took the other, as just as fair,
And having perhaps the better claim,
Because it was grassy and wanted wear;
Though as for that the passing there
Had worn them really about the same,

And both that morning equally lay
In leaves no step had trodden black.
Oh, I kept the first for another day!
Yet knowing how way leads to way,
I doubted if I should ever come back.

I shall be telling this with a sigh
Somewhere ages and ages hence:
Two roads diverged in a wood, and I—
I took the one less traveled by,
And that has made all the difference...

By Robert Frost

5) Solving the First Problem

"To see a world in a grain of sand
And a heaven in a wildflower,
Hold infinity in the palm of your hand
And eternity in an hour..." William Blake

5.1) Differential Change Ratios

My father graduated from Purdue University in 1934 with a freshly-minted degree in electrical engineering. I can still revive him in my thoughts via a conversation from many years ago. I then asked, "What is calculus?" His answer was; "Calculus, that's just dee x, dee y, dee-two x, dee-two y, and *dee y over dee x*. What one fool can do another fool can do!" He said it very fast, trying to imitate an auctioneer. And today, four decades later, I firmly believe that calculus did not present a major learning problem for my father. And the reason for this was the power of the differential as he so 'eloquently' expressed it above.

Old (pre 1930) calculus textbooks sometimes had high-sounding titles that incorporated the word *infinitesimal*. Infinitesimal, as stated in Chapter 4, is just another descriptive word for differential. Yet another word that early 20th century authors would use when describing a differential is *an indivisible*. For example, they would use the symbol dx and call the quantity represented *an indivisible* of x. The idea is that dx is so small that it cannot be partitioned (or divided) any further. The idea of extreme smallness—as expressed by *an indivisible*—is definitely on target, but the idea of not being able to subdivide further totally misses the mark. Remember my father's words: *dee y over dee x*. In mathematical terms, this rational quantity is written

$$\frac{dy}{dx}.$$

And, as the symbol plainly states, the above is simply dy divided by dx —one wee number divided by another wee number.

Now, when forming $\dfrac{dy}{dx}$ the two variables x and y are presumed to be in a functional relationship of the form $y = f(x)$ where the infinitesimal change dy has been induced by the infinitesimal change dx. The quantity $\dfrac{dy}{dx}$ is a *differential change ratio* and has the same units as any other change ratio studied thus far, which is units of y change per single unit of x change.

Just how big is a *differential change ratio* since differentials themselves are extremely small quantities? To answer, suppose we have a function $y = f(x)$ where $dx = .00000000001$ and the corresponding dy is calculated to be $dy = .000000000073$. Both of these differential quantities are measured in trillionths, which is undoubtedly small. Let's divide:

$$\frac{dy}{dx} = \frac{.000000000073}{.00000000001} = 7.3 \frac{units \cdot of \cdot y}{unit \cdot of \cdot x}.$$

Notice that simple division is a marvelous process that can manufacture numerically significant outcomes from two very small quantities. Here, two very small quantities have been divided, producing a respectable 7.3. The digit to the right of the decimal point is definite proof of precision divisibility at the micro-level. Let's make one final reinforcing point with this example. Think about the absolute magnitudes of $dx = .00000000001$ and $dy = .000000000073$. What could one say about the magnitude of $(dx)^2$, or $(dy)^2$, or $(dx)(dy)$, or $(dx)^3$, and so on? All of these higher-order quantities are so small that they have to be measured in terms of quintillionths or smaller. *Hence, these higher-order quantities don't even show up on the numerical radar screen.*

To compute $\dfrac{dy}{dx}$ from a functional relationship $y = f(x)$, first develop the expression $dy = g(x)dx$ as shown in Section 4.7. Next, divide both sides by the differential dx to obtain

$$\frac{dy}{dx} = g(x).$$

In the following example, this brute-force computational technique is applied to a moderately difficult problem where the *casting out* of higher-order differentials paves the way to the desired solution.

Ex 5.1.1: Find $\dfrac{dy}{dx}$ for $y = \sqrt[3]{(2x-5)^2}$ or $y^3 = (2x-5)^2$

$$y = \sqrt[3]{(2x-5)^2} \Rightarrow$$
$$y + dy = \sqrt[3]{(2[x+dx]-5)^2} \Rightarrow$$
$$(y+dy)^3 = (2[x+dx]-5)^2 \Rightarrow$$
$$y^3 + 3y^2 dy + 3y(dy)^2 + (dy)^3 = (2[x+dx]-5)^2 \Rightarrow$$
$$y^3 + 3y^2 dy = 4[x^2 + 2xdx + (dx)^2] - 20[x+dx] + 25 \Rightarrow$$
$$y^3 + 3y^2 dy = 4x^2 - 20x + 25 + [8x-20]dx \Rightarrow$$
$$y^3 + 3y^2 dy = (2x-5)^2 + [8x-20]dx \Rightarrow$$
$$3y^2 dy = [8x-20]dx \Rightarrow$$
$$\frac{dy}{dx} = \frac{8x-20}{3y^2} \Rightarrow \frac{dy}{dx} = \frac{8x-20}{3\sqrt[3]{(2x-5)^4}} \quad \therefore$$

Computational shortcuts, bypassing the brute-force approach shown above, will be developed in Section 5.3. The final result

$$\frac{dy}{dx} = \frac{8x-20}{3\sqrt[3]{(2x-5)^4}}$$

is a brand new function derived from $y = \sqrt[3]{(2x-5)^2}$. Notice that this new function is not defined at $x = \frac{5}{2}$, even though the original function is defined at the same point.

82

We are ready to present one of the major results in this book. Let $y = f(x)$. Recall the definition for the derivative $f'(x)$:

$$f'(x) = \lim_{h \to 0} \left[\frac{f(x+h) - f(x)}{h} \right].$$

Now $h \to 0$ implies that $\dfrac{f(x+h) - f(x)}{h} \to f'(x)$ since $f'(x)$ is the value of the limit. Now imagine that the limit process for manufacturing $f'(x)$ is currently 'in motion' and not yet been completed. This should bring to mind the image of an incredible shrinking h getting smaller by the second. Also, imagine that the dependent expression $\dfrac{f(x+h) - f(x)}{h}$ has begun to target $f'(x)$, eventually to merge with it as h continues on its rapid collapse towards zero. Suppose we suddenly put on the brakes and stop h just short of zero as shown in **Figure 5.1**. This saves h from a scheduled fate of sliding into oblivion.

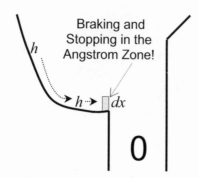

Figure 5.1: Saving h from Oblivion

How short is short you might ask? Short enough so that h for all effects and purposes is a dx. It is short enough so that the remaining closing distance between $\dfrac{f(x+h) - f(x)}{h}$ and $f'(x)$ is of no consequence.

Thus, once h has been brought to rest within the *numerical Angstrom Zone* depicted in **Figure 5.1** (*Note: one angstrom equals one ten millionth of a millimeter.*), we have that:

$$f'(x) = \frac{f(x+dx) - f(x)}{dx} = \frac{y + dy - y}{dx} = \frac{dy}{dx}.$$

The two ends of the above equality state that the derivative and differential change ratio are one and the same—an amazing result. Three immediate and interlinked consequences are

$$\frac{dy}{dx} = f'(x), \; dy = f'(x)dx, g(x) = f'(x).$$

Recall that Section 4.7 defines $g(x)$ to be the final result in functional form when developing the differential relationship $dy = g(x)dx$.

Ex 5.1.2: Verify that $\dfrac{dy}{dx} = f'(x)$ for $f(x) = y = x^2 - 3x - 4$.

By **Ex 4.6.1**, we have that $f'(x) = 2x - 3$.

$$y + dy = (x + dx)^2 - 3(x + dx) - 4 \Rightarrow$$
$$dy = (x + dx)^2 - 3(x + dx) - 4 - y \Rightarrow$$
$$dy = x^2 + 2xdx + (dx)^2 - 3x - 3dx - 4 - [x^2 - 3x - 4] \Rightarrow$$
$$dy = 2xdx + (dx)^2 - 3dx \Rightarrow dy = (2x - 3)dx \Rightarrow$$
$$\frac{dy}{dx} = 2x - 3 = f'(x) \therefore$$

Let's recap the previous sequence of events: h has been brought to rest in the Angstrom Zone; and, in this zone, the now-resting h has become small enough to be converted into a differential dx. As a result, the following functional equality also holds for any so-converted h in the Angstrom Zone:

$$\frac{f(x+dx)-f(x)}{dx} = f'(x) \Rightarrow f(x+dx) = f(x) + f'(x)dx$$

or

$$\frac{y+dy-y}{dx} = f'(x) \Rightarrow y+dy = y + f'(x)dx.$$

Figure 5.2 depicts a greatly magnified view—much like that seen through a modern scanning-electron microscope—of the graph of the function $y = f(x)$ between the two points (x, y) and $(x+dx, y+dy)$. This is the world of the Angstrom Zone where the function $y = f(x)$ is linear *for all computational purposes* between the two points (x, y) and $(x+dx, y+dy)$. The slope of the straight line connecting the pair is given by the value of the derivative $f'(x)$ at x.

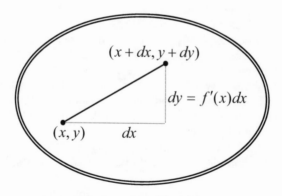

Figure 5.2: Greatly Magnified View of $y = f(x)$

Linearity of f in our conceptual micro-world is perfectly consistent with the notion that dx must be small enough so that all second order or higher (non-linear) effects can be totally ignored—of no consequence whatsoever. This *micro-linearity*, which allowed for an exact analysis of functionally linked change behavior at a given point $(x, y = f(x))$, was precisely the power of the differential as originally conceived by the super minds of Newton and Leibniz.

Returning to Barrow's Diagram (**Figure 2.1**), the small shaded triangle was called a *differential triangle*; and dx was presumed small enough so that $y = f(x)$ could be computationally treated as linear—having slope $f'(x)$—between the two endpoints of the triangle's hypotenuse. The lune-like area nestled between the curve and triangle, though visible in the representation below, was assumed to be virtually nonexistent.

Today, mathematicians can dream up and construct numerous functional situations where the differential possesses none of the classical linear properties as envisioned by Barrow, Newton and Leibniz. Many counter-examples can be found in the area of fractal geometry (e.g. the Mandelbrot set) where functional patterns replicate their intricacies ad-infinitum as one descends into the Lilliputian world. However, we shall not worry about fractals and their subsequent lack of linear behavior in the classical world of the small. For, it was the classical world of the small—as originally conceived by Newton, Leibniz, and others—that led to the modern discipline of celestial mechanics and the first lunar landing on 20 July 1969, a topic explored further in Chapter 8.

$$\int_{a}^{b} \overset{\cdot\cdot}{\cup} dx$$

Verify that $\dfrac{dy}{dx} = f'(x)$ for $y = f(x) = \dfrac{1}{x^2}$. Use the process

defined by $f'(x) = \lim\limits_{h \to 0}\left[\dfrac{\Delta y}{\Delta x}\right] = \lim\limits_{h \to 0}\left[\dfrac{f(x+h)-f(x)}{h}\right]$ to make

$f'(x)$. Use the differential process as defined by the relationship

$y + dy = f(x + dx)$ to make $\dfrac{dy}{dx}$. Which of these two processes

represents an easier path to the *common final product*?

5.2) Process and Products: Differentiation

The title phrase brings to mind visions of a manufacturing facility where workers assemble products for the modern consumer via a pre-determined sequence of steps (a process). This is certainly a correct understanding of what is meant by 'Process and Products', but it need not be the only understanding. Much of mathematics can also be thought of in terms of processes and associated products. Functions, as defined in Section 4.1, are great examples of processes and products where numerical input (the raw material) is being processed by a sequence of steps (usually algebraic in nature) in order to produce numerical output. Our newest example is *the process of differentiation, which produces products known as derivatives*.

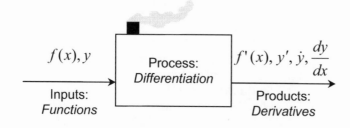

Figure 5.3: The Process of Differentiation

As **Figure 5.3** depicts, differentiation is the process by which we make, or derive, derivatives from functions. Inputs to the differentiation process are functions expressed by either $f(x)$ or y notation. The products, called *first derivatives or first derived functions*, can be denoted as $f'(x)$, or y', or $\dot{y}$, or $\dfrac{dy}{dx}$ where the hash mark is read *prime*. All four notations mean the same thing and refer to the same quantity. Typically, $f'(x)$ is used when we want to emphasize the derivative as a geometric slope and wish to compute, as a new function of x in its own right, specific values of $f'(x)$. The notation y' is the symbol of choice when derivatives, their associated independent variables, and parent functions appear together in an algebraically assembled equation known as a *differential equation* as shown below:

$$y' - xy = x^2 + 1.$$

Note: Starting with Section 6.4, we will explore the formulation, solution, and use of elementary differential equations, one of the major topics in this book. When the independent variable is time, the derivative equates to instantaneous velocity; customarily indicated by the *dot* notation $\dot{y}$. Finally, $\dfrac{dy}{dx}$ emphasizes the fact that the derivative is an instantaneous or differential change ratio for the two infinitesimal quantities dx and dy linked via a functional relationship of the form $y = f(x)$. Each of four notations can be used to signify both process and product. For example, the equation $(y)' = y'$ states that the product from the differentiation process $(y)'$ is the derivative y'.

Presently, there are two different methods by which we can conduct the differentiation process. One method utilizes limits, and the other method utilizes differentials. If properly carried out, both lead to the derivative being sought. However, neither method is without computational difficulty as the next example shows.

Ex 5.2.1: Differentiate $f(x) = y = \sqrt{x^2 + 4}$

Method 1: Use the process $f'(x) = \lim\limits_{h \to 0}\left[\dfrac{f(x+h) - f(x)}{h}\right]$.

Note: In the second and third lines below, the numerator is rationalized using standard algebraic techniques. This algebraic rearrangement allows the h in the denominator to be cancelled, which, in turn, creates a limit that can be determined.

$$f'(x) = \lim_{h \to 0}\left[\frac{\sqrt{(x+h)^2 + 4} - \sqrt{x^2 + 4}}{h}\right] \Rightarrow f'(x) =$$

$$\lim_{h \to 0}\left[\left(\frac{\sqrt{(x+h)^2 + 4} - \sqrt{x^2 + 4}}{h}\right)\left(\frac{\sqrt{(x+h)^2 + 4} + \sqrt{x^2 + 4}}{\sqrt{(x+h)^2 + 4} + \sqrt{x^2 + 4}}\right)\right] \Rightarrow$$

$$f'(x) = \lim_{h \to 0}\left[\frac{(x+h)^2 + 4 - \{x^2 + 4\}}{h(\sqrt{(x+h)^2 + 4} + \sqrt{x^2 + 4})}\right] \Rightarrow$$

$$f'(x) = \lim_{h \to 0}\left[\frac{x^2 + 2xh + h^2 + 4 - \{x^2 + 4\}}{h(\sqrt{(x+h)^2 + 4} + \sqrt{x^2 + 4})}\right] \Rightarrow$$

$$f'(x) = \lim_{h \to 0}\left[\frac{2xh + h^2}{h(\sqrt{(x+h)^2 + 4} + \sqrt{x^2 + 4})}\right] \Rightarrow$$

$$f'(x) = \lim_{h \to 0}\left[\frac{2x + h}{(\sqrt{(x+h)^2 + 4} + \sqrt{x^2 + 4})}\right] \Rightarrow$$

$$f'(x) = \frac{2x}{2\sqrt{x^2 + 4}} = \frac{x}{\sqrt{x^2 + 4}} \quad \therefore$$

Method 2: Use the process $y + dy = f(x + dx)$.

$$y + dy = f(x + dx) = \sqrt{(x + dx)^2 + 4} \Rightarrow$$
$$(y + dy)^2 = (x + dx)^2 + 4 \Rightarrow$$

$$y^2 + 2ydy + (dy)^2 = x^2 + 2xdx + (dx)^2 + 4 \Rightarrow$$
$$y^2 + 2ydy = (x^2 + 4) + 2xdx \Rightarrow$$
$$2ydy = 2xdx \Rightarrow$$
$$\frac{dy}{dx} = \frac{x}{y} = \frac{x}{\sqrt{x^2+4}}$$

As **Ex 5.2.1** shows, both methods lead to the same result, and both are somewhat cumbersome to execute. **Method 1** requires that the numerator be rationalized (via the use of some fairly sophisticated algebra) before the h in the denominator can be divided out, which is a precondition to taking the limit. **Method 2**, though much simpler, still requires the attentive use of differential fundamentals throughout the differentiation process.

In the quality world, we often talk about process improvement Product improvement is defined as any reduction in the number of processing steps that leads to time and/or cost savings in producing the product. The differentiation process, no matter which of the above methods is used, can be somewhat lengthy and challenging to execute as illustrated by **Ex 5.2.1**. Some major improvement is clearly desired and needed in the differentiation process. The next section, Section 5.3, will present several major process improvements. These improvements, typically algebraic in nature, will greatly streamline what pre-quality-era writers called *the taking of derivatives*.

$$\int_a^b \overset{\bullet\bullet}{\cup} dx$$

Section Exercise

Differentiate $y = x$, $y = x^2$, $y = x^3$, $y = x^4$, and $y = x^5$ using the differential method. Do you notice a general pattern? Now go ahead and differentiate $y = x^{143}$ *based on your experience*.

5.3) Process Improvement: Derivative Formulas

Derivatives are products, obtained from *parent functions* via the process of differentiation. This rather extensive section presents and illustrates thirteen powerful algebraic formulas that greatly speed up the differentiation process by *pattern matching*. In college algebra, pattern matching is the process used when numerical quantities from a specific quadratic equation are substituted into the general quadratic formula in order to produce a solution. Due to length, Section 5.3 is sub-sectioned into formula/pattern groupings. Each subsection will follow the general order: presentation of the formula, formula derivation(s), and formula illustration(s). All thirteen derivative formulas are repeated for reference in Appendix D.

Most of the formulas presented in this section will be derived using basic definitions and differentials. In this way, the power of the differential will be continually displayed. One notable exception is the function $y = e^x$, a case where differentials and limits are both used to develop the formula for *taking the derivative*. The binomial theorem is also extensively utilized in the derivation of several of the differentiation formulas. Hence, a brief review of this fundamental algebraic result is in order. The binomial theorem states that for a binomial expression $(x + y)$ raised to a positive integer power n, we have:

$$(x+y)^n = \sum_{i=0}^{n} \binom{n}{i} x^{n-i} y^i \text{ , where } \binom{n}{i} = \frac{n!}{i!(n-i)!} \text{ .}$$

The expression $\sum_{i=0}^{n} \binom{n}{i} x^{n-i} y^i$ is known as the *binomial expansion* for $(x + y)^n$. Each of the two terms inside the parenthesis can be any algebraic quantity whatsoever, simple or complicated. As we shall see, this makes the binomial expansion a very flexible and powerful tool for evaluating differential expressions of the form $(x + dx)^n$ or $(y + dy)^n$.

Ex 5.3.1: Evaluate the expressions $(x+dx)^4$ and $(y+dy)^6$ where the two differentials dx, dy are associated with an independent variable x and the associated dependent variable y.

By the binomial expansion, we have for $(x+dx)^4$

$$(x+dx)^4 = \sum_{i=0}^{4}\binom{4}{i}x^{4-i}(dx)^i =$$

$$x^4 + \frac{4!}{1!3!}x^3(dx)^1 + \frac{4!}{2!2!}x^2(dx)^2 + \frac{4!}{3!1!}x^1(dx)^3 + (dx)^4 =$$

$$x^4 + 4x^3(dx) + [6x^2(dx)^2 + 4x(dx)^3 + (dx)^4]$$

Now, the trinomial $6x^2(dx)^2 + 4x(dx)^3 + (dx)^4$ consists entirely of Higher Order Differential Terms ($HODT$), terms that are always totally negligible under the *Rules of Engagement* for differentials (Section 4.7). Hence our expansion reduces to

$$(x+dx)^4 = x^4 + 4x^3(dx) \therefore$$

Note: A word of caution is needed here. The binomial expansion is used throughout mathematics, not just in calculus and with differentials. Higher Order Differential Terms can be thrown out because each term contains an infinitesimally small factor being raised to a second-order power or better. This power-raising, for all practical purposes, makes the term disappear. In binomial expansions where both terms inside the parenthesis are of 'normal' magnitude, all higher order terms must be retained.

Continuing with $(y+dy)^6$:

$$(y+dy)^6 =$$
$$y^6 + 6y^5(dy) + 15y^4(dy)^2 + 20y^3(dy)^3 +$$
$$15y^2(dy)^4 + 6y(dy)^5 + (dy)^6 \Rightarrow$$
$$(y+dy)^6 = y^6 + 6y^5(dy) \therefore$$

5.3.1) Four Basic Differentiation Rules

The four basic differentiation rules can be evenly split into two categories: 1) Specific Rules for Special Functions, and 2) General Process Improvement Rules. **R1** and **R2** are Category 1 rules; **R3** and **R4**, Category 2 rules.

R1. Derivative of a Constant $k : \left[k\right]' = 0$

Proof: $y = f(x) = k \Rightarrow y + dy = k \Rightarrow dy = 0 \Rightarrow \dfrac{dy}{dx} = 0$

Illustration: $f(x) = 17 \Rightarrow f'(x) = 0$

Illustration: $y = \pi \Rightarrow \dfrac{dy}{dx} = 0$

R2. Basic Power Rule: $\left[x^n\right]' = nx^{n-1}$ where n can be any number whatsoever—positive, negative, rational, etc.

Proof for Positive Integers: binomial expansion used

$y = x^n \Rightarrow y + dy = (x + dx)^n \Rightarrow$

$y + dy = x^n + nx^{n-1}(dx) + HODT \Rightarrow$

$dy = nx^{n-1}(dx) \Rightarrow \dfrac{dy}{dx} = y' = nx^{n-1} \therefore$

Proof for Simple Radicals: binomial expansion used

$y = x^{\frac{1}{n}} \Rightarrow y + dy = (x + dx)^{\frac{1}{n}} \Rightarrow$

$(y + dy)^n = x + dx \Rightarrow y^n + ny^{n-1}dy + HODT = x + dx \Rightarrow$

$ny^{n-1}dy = dx \Rightarrow \dfrac{dy}{dx} = \dfrac{1}{ny^{n-1}} \Rightarrow \dfrac{dy}{dx} = \dfrac{1}{n(x^{\frac{1}{n}})^{n-1}} \Rightarrow$

$\dfrac{dy}{dx} = \dfrac{1}{nx^{1-\frac{1}{n}}} \Rightarrow \dfrac{dy}{dx} = y' = \dfrac{1}{n}x^{\frac{1}{n}-1} \therefore$

Proof for Negative Integers: binomial expansion used

$$y = x^{-n} \Rightarrow y + dy = (x + dx)^{-n} \Rightarrow y + dy = \frac{1}{(x + dx)^n} \Rightarrow$$

$$(y + dy)(x + dx)^n = 1 \Rightarrow (y + dy)(x^n + nx^{n-1}dx + HODT) = 1 \Rightarrow$$

$$(y + dy)(x^n + nx^{n-1}dx) = 1 \Rightarrow$$

$$yx^n + nyx^{n-1}(dx) + x^n dy + nx^{n-1}(dx)(dy) = 1 \Rightarrow$$

$$nyx^{n-1}dx + x^n dy = 0 \Rightarrow \frac{dy}{dx} = -\frac{nyx^{n-1}}{x^n} \Rightarrow$$

$$\frac{dy}{dx} = y' = -nx^{-n-1} \therefore$$

Illustration: $f(x) = x^{45} \Rightarrow f'(x) = 45x^{44}$

Illustration: $y = \frac{1}{x^5} = x^{-5} \Rightarrow \frac{dy}{dx} = y' = -5x^{-6} = -\frac{5}{x^6}$

Note: as shown above, all exponential expressions must be put in the general exponential form a^b before applying the power rule.

Illustration: $f(x) = \sqrt[9]{x} = x^{\frac{1}{9}} \Rightarrow f'(x) = \frac{1}{9} x^{\frac{1}{9}-1} = \frac{1}{9\sqrt[9]{x^8}}$

R3. Coefficient Rule: $\left[\alpha f\right]' = \alpha f'$ where α can be any numerical coefficient and $f = f(x)$

Proof: Let $w = \alpha f$

$$w + dw = \alpha(f + df) = \alpha f + \alpha df \Rightarrow dw = \alpha df \Rightarrow$$

$$\frac{dw}{dx} = \alpha \frac{df}{dx} \Rightarrow w' = \alpha[f'] \Rightarrow [\alpha f]' = \alpha f' \therefore$$

Note: The coefficient rule is our first true process-improvement rule. Instead of giving a specific result, it shows how to streamline the differentiation process for a general function f.

Illustration:
$$f(x) = 19x^9 \Rightarrow f'(x) = [19x^9]' \Rightarrow$$
$$f'(x) = 19[x^9]' = 19(9)x^8 = 171x^8$$

R4. Sum and Difference Rule: $[f \pm g]' = f' \pm g'$ where $f = f(x)$ and $g = g(x)$

Proof: Let $w = f \pm g$

$$w + dw = f + df \pm (g + dg) \Rightarrow w + dw = f \pm g + df \pm dg \Rightarrow$$
$$dw = df \pm dg \Rightarrow \frac{dw}{dx} = \frac{df}{dx} \pm \frac{dg}{dx} \Rightarrow w' = f' \pm g' \Rightarrow$$
$$[f \pm g]' = f' \pm g' \therefore$$

Note: The Sum and Difference rule is our second true process-improvement rule. The rule can be easily extended to the sum and difference of an arbitrary number of functions as the following illustration shows.

In the next illustration, the four basic differentiation rules work together in seamless functioning, allowing one to manufacture a complicated derivative without any direct use of limits or differentials. This improved process, totally algebraic in nature, relies heavily on your ability to *pattern-match*. Much of the art of taking derivatives is the knowing of how and when to apply the general differentiation rules to a particular function. This knowledge comes from concerted hands-on practice, and not from the observation of another's mathematical skills in action. My sincere advice for those who wish to become competent in differentiation is to spend much of your time in practice!

Illustration: Find y' for $y = 9x^4 - 2\sqrt{x} + \dfrac{6}{x^3} + 17$.

$$y = 9x^4 - 2\sqrt{x} + \frac{6}{x^3} + 17 \Rightarrow y = 9x^4 - 2x^{\frac{1}{2}} + 6x^{-3} + 17 \Rightarrow$$

$$y' = [9x^4 - 2x^{\frac{1}{2}} + 6x^{-3} + 17]' \Rightarrow$$

$$y' = [9x^4]' - [2x^{\frac{1}{2}}]' + [6x^{-3}]' + [17]' \Rightarrow$$

$$y' = 9[x^4]' - 2[x^{\frac{1}{2}}]' + 6[x^{-3}]' + [17]' \Rightarrow$$

$$y' = 9 \cdot 4x^3 - 2 \cdot \tfrac{1}{2} x^{-\frac{1}{2}} + 6 \cdot (-3)x^{-4} + 0 \Rightarrow$$

$$y = 36x^3 - \frac{1}{\sqrt{x}} - \frac{18}{x^4}$$

5.3.2) Five Advanced Differentiation Rules

R5: Product Rule: $[fg]' = fg' + gf'$ where $f = f(x)$ and $g = g(x)$. The product rule is our first counterintuitive differentiation rule in that $[fg]' \neq f'g'$. *In words, the derivative of a product is not equal to the product of the derivatives.*

Proof: Let $w = fg$

$$w + dw = (f + df)(g + dg) \Rightarrow$$
$$w + dw = fg + gdf + fdg + (df)(dg) \Rightarrow$$
$$dw = gdf + fdg + (df)(dg) \Rightarrow dw = gdf + fdg \Rightarrow$$
$$\frac{dw}{dx} = \frac{gdf + fdg}{dx} \Rightarrow \frac{dw}{dx} = g\frac{df}{dx} + f\frac{dg}{dx} \Rightarrow$$
$$w' = gf' + fg' \Rightarrow [fg]' = gf' + fg' \therefore$$

The correctness of the product rule $[fg]' = fg' + gf'$ is demonstrated in the illustration below, an illustration which also demonstrates the <u>incorrectness</u> of $[fg]' = f'g'$, an error common to calculus beginners.

Illustration: Find $[fg]'$ when $f(x) = x^9$ and $g(x) = x^5$.

Correct: $y = x^9 x^5 = x^{14} \Rightarrow y' = 14x^{13}$ our benchmark.

Correct:
$$y = x^9 x^5 \Rightarrow y' = [x^9][x^5]' + [x^5][x^9]' \Rightarrow$$
$$y' = [x^9]5x^4 + [x^5]9x^8 \Rightarrow y' = 5x^{13} + 9x^{13} = 14x^{13}$$

Incorrect:
$$y = x^9 x^5 \Rightarrow y' = [x^9]'[x^5]' \Rightarrow$$
$$y' = [9x^8][5x^4] \Rightarrow y' = 45x^{12} \neq 14x^{13}$$

Figure 5.4 (akin to **Figure 4.14**) depicts the differential change relationship for a product of two functions $w = fg$. The four-piece rectangle represents $w + dw = (f + df)(g + dg)$, and the one non-shaded rectangle represents $w = fg$. The two lightly shaded rectangles collectively represent the infinitesimal change dw in w corresponding to an infinitesimal change dx in the independent variable x. The one darker rectangle is second order, and is, *of course*, totally negligible, leaving us with $dw = gdf + fdg$.

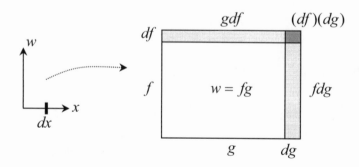

Figure 5.4: Differential Change Relationship for $w = fg$

R6: Quotient Rule: $\left[\dfrac{f}{g}\right]' = \dfrac{gf' - fg'}{g^2}$ where $f = f(x)$ and $g = g(x)$. The quotient rule is our second counterintuitive differentiation rule in that $\left[\dfrac{f}{g}\right]' \neq \dfrac{f'}{g'}$.

Proof: Let $w = \dfrac{f}{g}$

$$w + dw = \frac{f + df}{g + dg} \Rightarrow w + dw = \frac{(f + df)(g - dg)}{(g + dg)(g - dg)} \Rightarrow$$

$$w + dw = \frac{fg + gdf - fdg - (df)(dg)}{g^2 - (dg)^2} \Rightarrow$$

$$w + dw = \frac{fg + gdf - fdg}{g^2} \Rightarrow$$

$$w + dw = \frac{fg}{g^2} + \frac{gdf - fdg}{g^2} \Rightarrow dw = \frac{gdf - fdg}{g^2} \Rightarrow$$

$$\frac{dw}{dx} = \frac{\dfrac{gdf - fdg}{dx}}{g^2} \Rightarrow \frac{dw}{dx} = \frac{g\dfrac{df}{dx} - f\dfrac{dg}{dx}}{g^2} \Rightarrow$$

$$w' = \frac{gf' - fg'}{g^2} \Rightarrow [fg]' = \frac{gf' - fg'}{g^2} \quad \therefore$$

As we did with the product rule, the correctness of the quotient rule $\left[\dfrac{f}{g}\right]' = \dfrac{gf' - fg'}{g^2}$ is demonstrated below, as well as the <u>incorrectness</u> of $\left[\dfrac{f}{g}\right]' = \dfrac{f'}{g'}$, another common error.

Illustration: Find $\left[\dfrac{f}{g}\right]'$ when $f(x) = x^9$ and $g(x) = x^5$.

Correct: $y = \dfrac{x^9}{x^5} = x^4 \Rightarrow y' = 4x^3$ our benchmark.

$$y = \frac{x^9}{x^5} \Rightarrow y' = \frac{[x^5][x^9]' - [x^9][x^5]'}{(x^5)^2} \Rightarrow$$

Correct: $y' = \dfrac{[x^5]9x^8 - [x^9]5x^4}{x^{10}} \Rightarrow$

$$y' = \frac{9x^{13} - 5x^{13}}{x^{10}} = \frac{4x^{13}}{x^{10}} = 4x^3$$

$$y = \frac{x^9}{x^5} \Rightarrow y' = \frac{[x^9]'}{[x^5]'} \Rightarrow$$

Incorrect:

$$y' = \frac{9x^8}{5x^4} = \frac{9}{5}x^4 \neq 4x^3$$

R7: Chain Rule for Composite Functions: $[f(g)]' = f'(g)g'$ where $f = f(x)$, $g = g(x)$ and $f(g) = f(g(x))$.

Proof: Let $w = f(g)$ and recall that $dg = g'dx$

$dw = f'(g)dg \Rightarrow dw = f'(g)\{g'dx\} \Rightarrow$
$dw = f'(g)g'dx \Rightarrow$
$\dfrac{dw}{dx} = f'(g)g' \Rightarrow w' = [f(g)]' = f'(g)g'$ ∴.

Although simple to prove if using differentials, the chain rule is not that simple to use. Pattern recognition, gained only through practice, is very definitely the key to success as shown in the following illustration.

Illustration: Find y' for $y = (x^2 + 1)^{10}$.

Notice that y can be thought of in terms of the composite function $f(g(x))$ where $g(x) = x^2 + 1$ and $f(x) = x^{10}$.   Proceeding   with the differentiation process, we have:

$$y' = [(x^2 + 1)^{10}]' = 10(x^2 + 1)^9 \{x^2 + 1\}' \Rightarrow$$
$$y' = 10(x^2 + 1)^9 2x \Rightarrow y' = 20x(x^2 + 1)^9$$

Illustration: Find y' for $y = x\sqrt{x^2 + 1}$

$$y = x\sqrt{x^2 + 1} \Rightarrow y = x(x^2 + 1)^{\frac{1}{2}} \Rightarrow$$
$$y' = x[(x^2 + 1)^{\frac{1}{2}}]' + [x]'(x^2 + 1)^{\frac{1}{2}} \Rightarrow$$
$$y' = x[(\tfrac{1}{2})(x^2 + 1)^{-\frac{1}{2}}(2x)] + [1](x^2 + 1)^{\frac{1}{2}} \Rightarrow$$
$$y' = \frac{x^2}{(x^2 + 1)^{\frac{1}{2}}} + (x^2 + 1)^{\frac{1}{2}} \Rightarrow$$
$$y' = \frac{2x^2 + 1}{\sqrt{x^2 + 1}}$$

Note: In the above illustration, the product rule, chain rule, and three basic rules were all used in concert to conduct the differentiation process.

Without a doubt, the chain rule is the single most powerful differentiation rule in calculus. It is actually used to prove the next differentiation rule in our list, called the inverse rule.

R8: Inverse Rule for Inverse Functions: $[f^{-1}(x)]' = \dfrac{1}{f'(f^{-1}(x))}$

where the function $f = f(x)$ is $f_{1 \to 1}$.

Proof: Recall that $f(f^{-1}(x)) = x$

$$[f(f^{-1}(x))]' = [x]' \Rightarrow f'(f^{-1}(x))[f^{-1}(x)]' = 1 \Rightarrow$$
$$[f^{-1}(x)]' = \frac{1}{f'(f^{-1}(x))} \ \therefore$$

In words, the final result states that the derivative of an inverse function is the reciprocal of the derivative of the *forward* function composed with the inverse function proper. The inverse rule is frequently used to develop the differentiation formulas for inverse transcendental (non-algebraic) functions. We will touch upon this method in the next subsection. In the meantime, we will reprove the differentiation formula for simple radical exponents via the inverse rule.

Illustration: Use the inverse rule to show that $[\sqrt[n]{x}]' = \dfrac{1}{n\sqrt[n]{x^{n-1}}}$.

Let $f^{-1}(x) = \sqrt[n]{x}$ and $f_{1 \to 1}(x) = x^{n}$

$$[f^{-1}(x)]' = \frac{1}{n(f^{-1}(x))^{n-1}} \Rightarrow$$

$$[\sqrt[n]{x}]' = \frac{1}{n(\sqrt[n]{x})^{n-1}} \Rightarrow$$

$$[\sqrt[n]{x}]' = \frac{1}{n\sqrt[n]{x^{n-1}}}$$

The Generalized Power Rule, a special case of the Chain Rule, is especially useful in that it allows the user to quickly break down complicated functions having expressions raised to powers.

R9: Generalized Power Rule: $\left[(f)^{n}\right]' = n(f)^{n-1}f'$ where again, n can be <u>any exponent</u>.

Proof: *left to the reader…*

We will close this subsection with a comprehensive example that brings most of the differentiation rules given thus far into play. The reader might want to identify the rules used at each stage in the differentiation process.

Ex 5.3.2: Find y' for $y = \dfrac{2x^3\sqrt{2x^2+1}}{(x+1)^5} = \dfrac{2x^3(2x^2+1)^{\frac{1}{2}}}{(x+1)^5}$

$$y' = \left[\frac{2x^3(2x^2+1)^{\frac{1}{2}}}{(x+1)^5}\right]' \Rightarrow$$

$$y' = \frac{(x+1)^5[2x^3(2x^2+1)^{\frac{1}{2}}]' - 2x^3(2x^2+1)^{\frac{1}{2}}[(x+1)^5]'}{[(x+1)^5]^2} \Rightarrow$$

$$y' = \frac{(x+1)^5[2x^3(2x^2+1)^{\frac{1}{2}}]' - 2x^3(2x^2+1)^{\frac{1}{2}}[5(x+1)^4]}{(x+1)^{10}} \Rightarrow$$

$$y' = \frac{(x+1)[2x^3(2x^2+1)^{\frac{1}{2}}]' - 10x^3(2x^2+1)^{\frac{1}{2}}}{(x+1)^6} \Rightarrow$$

$$y' = \frac{(x+1)[6x^2(2x^2+1)^{\frac{1}{2}} + 4x^4(2x^2+1)^{-\frac{1}{2}}] - 10x^3(2x^2+1)^{\frac{1}{2}}}{(x+1)^6} \Rightarrow$$

$$y' = \frac{(x+1)[6x^2(2x^2+1) + 4x^4] - 10x^3(2x^2+1)}{(2x^2+1)^{\frac{1}{2}}(x+1)^6} \Rightarrow$$

$$y' = \frac{(x+1)[16x^4 + 6x^2] - 20x^5 - 10x^3}{(2x^2+1)^{\frac{1}{2}}(x+1)^6} \Rightarrow$$

$$y' = \frac{-4x^5 + 16x^4 - 4x^3 + 6x^2}{(2x^2+1)^{\frac{1}{2}}(x+1)^6} = \frac{2x^2(3 - 2x + 8x^2 - 2x^3)}{(2x^2+1)^{\frac{1}{2}}(x+1)^6}$$

Example 5.3.2 amply reinforces the old teacher's proverb: *You really learn algebra when you take calculus or calculus takes you.*

5.3.3) Four Differentiation Rules for Two Transcendental Functions

In this subsection, we are going to develop derivatives for the two transcendental functions $f(x) = e^x$ and $g(x) = \ln(x)$. For $f(x) = e^x$, first introduced in Section 4.3, both differentials and limits will be used to find $f'(x)$. For $g(x) = \ln(x)$, we will use the fact that $f(x)$ and $g(x)$ are inverse functions in order to find $g'(x)$. **Ex 5.3.3** below establishes that $g = f^{-1}$.

Ex 5.3.3: Show that $f^{-1}(x) = \ln(x)$ when $f(x) = e^x$.

For any logarithmic function to any positive base b, we have that $\log_b(a) = w \Rightarrow b^w = a$ by definition. Recall from college algebra that $\ln(x)$ is defined as $\log_e(x)$, which implies $\ln(x) = w \Rightarrow e^w = x$. In words, the output from the function $\ln(x)$ is the *exponent* that is placed on e in order to make the input x. Now, let's follow the action for the two function compositions $f(g(x))$ and $g(f(x))$ when $f(x) = e^x$ and $g(x) = \ln(x)$. For the function g to qualify as f^{-1}, we must have that $f(g(x)) = g(f(x)) = x$. Checking it out both ways:

$$f(g(x)) = e^{\ln(x)} = e^w = x$$
$$g(f(x)) = \ln(e^x) = \log_e(e^x) = x$$

Since $f(x) = e^x \Rightarrow f_{1 \to 1}$ and $g(x) = \ln(x)$ is exhibiting all the relational properties needed by f^{-1}, $f^{-1}(x) = \ln(x) \therefore$.

R10: Exponential Rule, base e : $\left[e^x \right]' = e^x$

Proof: $y + dy = e^{x+dx} = e^x e^{dx}$

In the next step, we are going to merge the limit and differential concepts in order to get an equivalent *binomial expression* for e^{dx}. Follow closely, for this is one of the more intricate developments in the book.

> *To take an exponential*
> *With a limit and differential*
> *Is a matter of grit,*
> *Persistence, and sweat*
> *When using just paper and pencil.*

June 2003

Recall $e^{rt} = \lim_{n \to \infty}\left[(1 + \tfrac{r}{n})^{nt}\right]$.

$$t = 1, r = dx \Rightarrow e^{dx} = \lim_{n \to \infty}\left[(1 + \tfrac{dx}{n})^{n}\right] \Rightarrow$$

$$e^{dx} = \lim_{n \to \infty}\left[\sum_{i=0}^{n}\binom{n}{i}(\tfrac{dx}{n})^{i}\right] \Rightarrow e^{dx} = \lim_{n \to \infty}\left[\sum_{i=0}^{n}\frac{n!}{i!(n-i)!}(\tfrac{dx}{n})^{i}\right] \Rightarrow$$

$$e^{dx} = \lim_{n \to \infty}\left[\sum_{i=0}^{n}\frac{n!}{i!(n-i)!}(\tfrac{dx}{n})^{i}\right] \Rightarrow$$

$$e^{dx} = \lim_{n \to \infty}\left[\sum_{i=0}^{n}\frac{n(n-1)(n-2)...(n-i+1)}{i!\,n^{i}}(dx)^{i}\right] \Rightarrow$$

$$e^{dx} = 1 + \frac{n}{n}(dx) +$$

$$\lim_{n \to \infty}\left[\sum_{i=2}^{n}\left\{\left(\frac{1}{i!}\right)\left(\frac{n}{n}\right)\left(\frac{n-1}{n}\right)\left(\frac{n-2}{n}\right)...\left(\frac{n-i+1}{n}\right)\right\}(dx)^{i}\right] \Rightarrow$$

$$e^{dx} = 1 + \frac{n}{n}(dx) + HODT \Rightarrow e^{dx} = 1 + dx$$

Finally:

$$y + dy = e^{x}e^{dx} \Rightarrow y + dy = e^{x}(1 + dx) \Rightarrow$$

$$dy = e^{x}dx \Rightarrow \frac{dy}{dx} = y' = e^{x} \;\therefore$$

Illustration: Find y' for $y = 7e^x + 2x$

$$y' = [7e^x + 2x]' \Rightarrow y' = 7[e^x]' + [2x]' \Rightarrow$$
$$y' = 7e^x + 2$$

Illustration: Find y' for $y = x^2 e^x$

$$y' = [x^2 e^x]' \Rightarrow y' = x^2 [e^x]' + e^x [x^2]' \Rightarrow$$
$$y' = x^2 e^x + 2xe^x \Rightarrow y' = x(x+2)e^x$$

Note: as illustrated above, the differentiation rule for $y = e^x$ operates in harmony with all other differentiation rules.

R11: General Exponential Rule, base $e : \left[e^{f(x)} \right]' = f'(x)e^{f(x)}$

Proof: Direct application of **R7**.

Illustration: Find y' for $y = 9e^{x^2+2x}$

$$y' = [9e^{x^2+2x}]' \Rightarrow y' = 9[e^{x^2+2x}]' \Rightarrow$$
$$y' = 9[x^2 + 2x]'e^{x^2+2x} \Rightarrow y' = 18(x+1)e^{x^2+2x}$$

R12: Logarithm Rule, base $e : \left[\ln(x) \right]' = \dfrac{1}{x}$

Proof: Use chain rule **R7** with $f(x) = e^x$, $f^{-1}(x) = \ln(x)$

$$e^{\ln(x)} = x \Rightarrow [e^{\ln(x)}]' = [x]' \Rightarrow$$
$$e^{\ln(x)}[\ln(x)]' = 1 \Rightarrow x[\ln(x)]' = 1 \Rightarrow$$
$$[\ln(x)]' = \frac{1}{x} \therefore$$

Illustration: Find y' for $y = \ln(x)e^x$

$$y' = [\ln(x)e^x]' \Rightarrow y' = \ln(x)[e^x]' + e^x[\ln(x)]' \Rightarrow$$

$$y' = \ln(x)e^x + \frac{e^x}{x} \Rightarrow y' = \left[\frac{x\ln(x)+1}{x}\right]e^x$$

R13: General Logarithm Rule, base e: $\left[\ln\{f(x)\}\right]' = \dfrac{f'(x)}{f(x)}$

Proof: Direct application of **R7**.

Illustration: Find y' for $y = 9\ln(x^2+1)$

$$y' = [9\ln(x^2+1)]' \Rightarrow y' = 9[\ln(x^2+1)]' \Rightarrow$$

$$y' = 9\left[\frac{2x}{x^2+1}\right] \Rightarrow y' = \frac{18x}{x^2+1}$$

Note: Don't assume y is algebraic in form just because y' is algebraic in form. Two immediate counterexamples are the functions $y = \ln(x)$ and $y = \ln(f(x))$ where $f(x)$ can be any rational function.

Ex 5.3.4: Our final big audacious example in this section uses most of the thirteen differentiation rules. Again, success in differentiation is knowing how and when to use the rules— knowledge achieved only through determined practice.

Find y' for $y = \dfrac{4x^2 e^{x^2}}{\ln(x)}$

$$y' = \left[\frac{4x^2 e^{x^2}}{\ln(x)}\right]' = \frac{\ln(x)[4x^2 e^{x^2}]' - 4x^2 e^{x^2}[\ln(x)]'}{[\ln(x)]^2} \Rightarrow$$

$$y' = \left[\frac{4x^2 e^{x^2}}{\ln(x)}\right]' \Rightarrow$$

$$y' = \frac{\ln(x)[4x^2 e^{x^2}]' - 4x^2 e^{x^2}[\ln(x)]'}{[\ln(x)]^2} \Rightarrow$$

$$y' = \frac{\ln(x)[8xe^{x^2} + 8x^3 e^{x^2}] - \dfrac{4x^2 e^{x^2}}{x}}{[\ln(x)]^2} \Rightarrow$$

$$y' = \frac{\ln(x)[8xe^{x^2} + 8x^3 e^{x^2}] - 4xe^{x^2}}{[\ln(x)]^2} \Rightarrow$$

$$y' = \frac{\ln(x)[8xe^{x^2} + 8x^3 e^{x^2}] - 4xe^{x^2}}{[\ln(x)]^2} \Rightarrow$$

$$y' = \frac{4xe^{x^2}\left[2x^2 \ln(x) + 2\ln(x) - 1\right]}{[\ln(x)]^2}$$

$$\int_a^b \smile dx$$

Section Exercises

Differentiate the following functions:

1) $y = 7x^2 - 4x + 2e^x + 17$

2) $y = \dfrac{x^2\sqrt{2x^2+1}}{2x+1}$

3) $y = x^3 \ln(x^2+1)$

4) $y = 4x^3 - 12x$

5) $f(x) = (x^4+1)^{12} e^{4x}$

6) $y = \dfrac{\ln(x)}{e^x}$

5.4) Applications of the Derivative

Now that we have an efficient means to produce the product f' from a given function f, is this product useful? The answer is an absolute yes. Not only will the derivative enable us to quite handily solve the First Fundamental Problem of Calculus set forth in Chapter 3, but it will also allow us to solve a variety of other problems that require a much deeper analysis of functions than algebra alone can provide.

Six generic applications of the derivative will be explored in Section 5.4. Each application is supported by one or more examples. The six applications by no means represent all possible applications, only a sampling of those which are more basic and foundational. Again, due to length, this Section is portioned into subsections.

Note: The power that Newton and Leibniz brought to the analysis of functions via the invention of the derivative is equivalent to the power that Galileo brought to the analysis of the universe via the invention of the telescope. Today, Galileo's universe is also being examined with mathematical functions, which are analyzed, in part, using the derivative as invented by Newton and Leibniz.

5.4.1) Tangent Lines and Normal Lines

<u>Definition</u>: Let f be a differentiable (able to manufacture the derivative) function at a point x_0. Then f has both a tangent line and normal line at the point x_0. These lines are defined by the two equations:

$$\text{Tangent: } y - f(x_0) = f'(x_0)(x - x_0)$$

$$\text{Normal: } y - f(x_0) = \frac{-1}{f'(x_0)}(x - x_0).$$

If $f'(x_0) = 0$, the two equations reduce to:

$$\text{Tangent: } y = f(x_0) + 0 \cdot x$$

$$\text{Normal: } x = x_0 + 0 \cdot y.$$

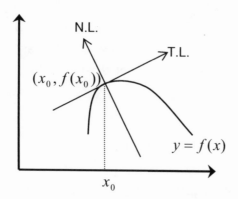

Figure 5.5: Tangent and Normal Lines

As **Figure 5.5** shows, the tangent line (T.L.) is that line which passes through the point $(x_0, f(x_0))$ and whose slope is identical to the slope of the function $f(x)$ at x_0. Hence, now that we have solved the First Fundamental Problem of Calculus, the slope of the tangent line is $f'(x_0)$. The normal line (N.L.) is that line which passes through the point $(x_0, f(x_0))$ and is perpendicular to the tangent line. By elementary analytic geometry, if two lines are perpendicular, then their two slopes m_1 and m_2 must satisfy the reciprocity relationship $m_1 = \dfrac{-1}{m_2}$, from which we can deduce that, the slope of the normal line is $\dfrac{-1}{f'(x_0)}$.

One can view the normal line as the path offering the quickest exit away from the graph of $y = f(x)$ at the point x_0. In **Figure 5.5**, pretend that the graph of $y = f(x)$ forms the upper boundary for a hot surface and that you are located at the point $(x_0, f(x_0))$. By moving away from the surface in the direction given by the normal line, you get the fastest cooling possible.

Ex 5.4.1: Given $y = f(x) = x^2 e^x$, find the equations of the tangent and normal lines at $x = 1$.

By the process-improvement formulas in Section 5.3, we have that $y' = f'(x) = 2xe^x + x^2 e^x$. Since both tangent and normal line equations require evaluation of f and f' at $x = 1$, substitute $x = 1$ to obtain $f(1) = 1^2 e^1 = e$ and $f'(1) = 3e$. Completing the equations, one obtains $y - e = 3e(x - 1)$ for the tangent line and $y - e = \dfrac{-1}{3e}(x - 1)$ for the normal line.

Ex 5.4.2: Find the equations of the tangent and normal lines for the function $y = f(x) = x^2 - 6x$ at the point $x = 4$. Repeat for any other x value(s) where we have $f'(x) = 0$. In this example, we also start using the symbol $\mapsto$ as defined in Chapter 1.

Where $x = 4$:
$$f(x) = x^2 - 6x \Rightarrow f'(x) = 2x - 6 \mapsto$$
$$f(4) = -8, f'(4) = 2 \mapsto$$
$$y + 8 = 2(x - 4) : TL$$
$$y + 8 = \tfrac{-1}{2}(x - 4) : NL$$

Where $f'(x) = 0$:
$$f'(x) = 2x - 6 = 0 \Rightarrow x = 3 \mapsto$$
$$f(3) = -9, f'(3) = 0 \mapsto$$
$$y = -9 + 0 \cdot x : TL$$
$$x = 3 + 0 \cdot y : NL$$

The slope of zero at $x = 3$ corresponds to a horizontal or level tangent line running parallel to the x axis. Hence, the normal line is vertical—straight up and down—running parallel to the y axis.

110

5.4.2) Newton's Method and Linear Approximation

Newton's method is an approximation method for finding solutions x to equations of the form $E(x) = 0$ or $E(x) = a$ where a is a real number. The expression $E(x)$ can be algebraic, transcendental, or combination thereof. Two such expressions are $x^3 + 2x + 1 = 0$ and $e^x - x^2 = 6$. Newton's method is a very simple example of a *numerical technique*. Numerical techniques in general are powerful mathematical *number-crunching* methods. They are used most often in conjunction with modern high-speed computers to solve equations that can't be solved in terms of a tidy algebraic and/or transcendental expression. *Numerical analysis* is the discipline where numerical techniques are studied in depth.

The first step in applying Newton's method is to recast the equations as functions, either $f(x) = E(x)$ or $f(x) = E(x) - a$. Hence, solving either one of the above equations is equivalent to finding the x intercepts for the function f.

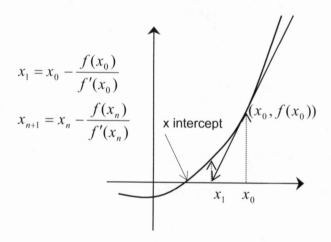

$$x_1 = x_0 - \frac{f(x_0)}{f'(x_0)}$$

$$x_{n+1} = x_n - \frac{f(x_n)}{f'(x_n)}$$

x intercept $(x_0, f(x_0))$

x_1 x_0

Figure 5.6: Schematic for Newton's Method

As depicted by **Figure 5.6**, the goal when applying Newton's Method is to find an approximate value for the desired x intercept.

The process is rather simple. First, pick a value x_0 known to be somewhat close to the intercept. Next, formulate the equation of the tangent line at x_0: $y - f(x_0) = f'(x_0)[x - x_0]$. The approach is to use the x intercept of the tangent line (easy to obtain) as an approximation for the x intercept of the function f (hard to obtain). Now set $y = 0$ in the above equation to obtain

$$-f(x_0) = f'(x_0)[x - x_0] \Rightarrow$$
$$x = x_0 - \frac{f(x_0)}{f'(x_0)} \quad or \quad x_1 = x_0 - \frac{f(x_0)}{f'(x_0)}$$

By **Figure 5.6**, x_1 is closer to the actual x intercept and becomes a new starting point for an even better approximation as the process is repeated as many times as necessary in order to achieve the desired accuracy.

Note: sometimes Newton's Method, as with any numerical technique, fails to converge—i.e. fails to come closer and closer to the desired solution— as the process continues to cycle itself. Convergence criteria (in terms of conditions on derivatives, etc.) exist for Newton's Method and for most other numerical methods. However, a discussion of convergence criteria is way beyond the scope of this primer. The 'new-fashion', practical way to test for convergence is to let the computer run the process and declare success if more and more digits are stabilized to the right of the decimal point. Likewise, if the computer crashes or gets in an endless do-loop, we definitely know that our particular problem did not converge. User judgment and experience becomes the deciding factor in either case.

The following example illustrates the use of Newton's Method.

Ex 5.4.3: Find the negative real zero for $x^3 + 2x + 1 = 0$.

Let $f(x) = x^3 + 2x + 1, f'(x) = 3x^2 + 2$. By the continuity discussion in Section 4.4, a simple polynomial function such as $f(x) = x^3 + 2x + 1$ does not have a break in its graph. Hence, one can claim that there is an x-axis crossover point (x intercept) in the interval $[-1, 0]$ since $f(-1) = -2$ and $f(0) = 1$.

Newton's Method, an iterative process, needs to be primed (like a pump) with a reasonable first guess or choice. Pick $x_0 = 0$ and begin.

$$x_1 = x_0 - \frac{f(x_0)}{f'(x_0)} = 0 - \frac{1}{2} = -.5 \Rightarrow$$

$$x_2 = -.5 - \frac{-.125}{2.75} = -.4545 \Rightarrow$$

$$x_3 = -.4545 - \frac{-.00288}{2.6197} = -.4534$$

We could continue with the process and expect to stabilize increasingly insignificant digits as the cycles repeat. The hundredths place, by all reasonable appearances, has been stabilized with three cycles. With an additional cycle, one could expect the thousandths place to be stabilized and so on. Notice that $f(.4545) = -.00288$. Remember that the goal is to find the x intercept, i.e. that value of x where $f(x) = 0$. Checking x_3, we have that $f(-.4534) = -.00000615$. Since x_3 results in a functional or output value which is only six millionths away from zero, we *choose* to stop the process via a *judgment call*.

Newton's method works very well for polynomial functions and converges quite rapidly when this is the case. For other classes of functions (e.g. exponential or logarithmic), Newton's Method can be a bit stubborn and take quite a few iterations to converge. Or, Newton's Method could fail to converge. User experience definitely counts.

We will finish this subsection with a discussion of simple linear approximation as given by the formula:

Linear Approximation Formula

$$f(x_1) \cong f(x_0) + f'(x_0)Dx : Dx \equiv x_1 - x_0$$

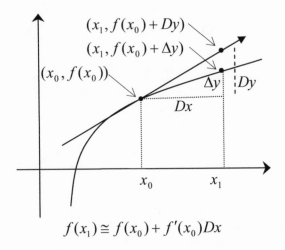

$$f(x_1) \cong f(x_0) + f'(x_0)Dx$$

Figure 5.7: The Basis of Linear Approximation

In **Figure 5.7**, the depicted tangent line has the equation $y - f(x_0) = f(x_0)(x - x_0)$. Let $x = x_1$ and y_1 let be the associated y value. Define the *macro quantity* Dx to be $x_1 - x_0$ and the macro quantity Dy to be $y_1 - f(x_0)$. Hence, we can rewrite the tangent line equation as $Dy = f'(x_0)Dx$.

Note: This is one of the few places in the book where I deviate from standard usage. When discussing approximation methods, many authors will reemploy the symbols dx and dy to represent the quantities Dx and Dy. This is since $\dfrac{Dy}{Dx} = f'(x_0) = \dfrac{dy}{dx}$. But, even though the above change ratios are identical, the actual D & d quantities are a universe apart. Recall that dx, dy are infinitesimals, while Dx, Dy are at least big enough to be seen with the naked eye on graph paper.

Continuing the discussion $\Delta y = f(x_1) - f(x_0)$ is the actual change in the function f when moving from x_0 to x_1.

As shown by **Figure 5.7** $\Delta y \neq Dy$, but there are many situations where $\Delta y \cong Dy$. The error, given by $|Dy - \Delta y|$, shrinks as Dx shrinks, which leads to the common-sense rule: *a smaller Dx is definitely more prudent*. Putting the pieces together, we have:

$$\Delta y \cong Dy = f'(x_0)Dx \Rightarrow$$
$$f(x_1) - f(x_0) \cong f'(x_0)Dx \Rightarrow .$$
$$f(x_1) \cong f(x_0) + f'(x_0)Dx$$

The last line is the linear approximation formula as previously blocked in gray. Two examples follow.

Ex 5.4.4: Approximate $\sqrt{27}$ using linear approximation.

Define $f(x) = \sqrt{x}$. The old trick is to pick the perfect square closest to 27, in this case 25. Set $x_1 = 27, x_0 = 25$ which implies $Dx = 2$. One quickly obtains $\sqrt{27} \cong \sqrt{25} + \dfrac{2}{2\sqrt{25}} = 5.2$.

Compare this to the actual value of 5.196.

Ex 5.4.5: Find the approximate change in volume of a big snowball when the radius increases from 2 feet to 2.1 feet.

Define $V(r) = \dfrac{4}{3}\pi r^3$. Hence $DV = 4\pi r^2 Dr$. Substituting $r = 2$ and $Dr = .1$, we have $DV = 5.0265$. Again, compare this to the actual value of $\Delta V = 5.282$. The 5% difference is probably pushing the upper limits of a reasonable quick approximation.

Note: In the days before large mainframes, simple hand-approximation techniques were indispensable to the practicing scientist and engineer in terms of time/labor saved. But as in Poker, you had to know 'when to fold them or when to hold them'—knowledge always gained through experience. It was this experience that lead to expert familiarity with the behavior of functions.

5.4.3) Finding Local Extrema

Chapter 5 is the longest and most extensive in the book, and Section 5.4 is the longest and most extensive in Chapter 5. Chapter 5 can be likened to a climb of K2, with the climb of Mount Everest to come later. One of the 'K2' topics that completely amazed me when first encountered was that of finding *local extrema*—to reveal itself as being exclusively in the realm of differential calculus, well beyond the reach of ordinary algebra.

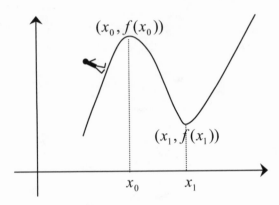

Figure 5.8: Local Maximum and Local Minimum

Let f be a function having a high point $(x_0, f(x_0))$ and a low point $(x_1, f(x_1))$ as shown in **Figure 5.8**.

Definition: The function f is said to have a *local maximum* at the point $(x_0, f(x_0))$ if there exists an interval $(x_0 - h, x_0 + h)$ centered at x_0 such that for all x in $(x_0 - h, x_0 + h)$ we have the relationship $f(x) < f(x_0)$. Exactly the opposite relationship exists, i.e. $f(x) > f(x_0)$, if we have a *local minimum*.

Translating, a local maximum is the highest point in a *surrounding* interval of points; and a local minimum, the lowest point in a surrounding interval of points. The key word is local: a local maximum may be akin to the tallest building in a small town.

116

Depending on the size of the town, the tallest building may or may be significant in terms of overall global stature (i.e. the tallest building in Dayton, Ohio is insignificant when compared to the tallest building in Chicago, Illinois). Looking again at **Figure 5.8**, one sees that there are points both higher and lower than $(x_0, f(x_0))$ and $(x_1, f(x_1))$ on the graph of f.

Now we are ready to present a MAJOR result addressing *local extrema*, the collective name for local maxima and minima.

Let f be differentiable in the interval $(x_0 - h, x_0 + h)$ and Suppose f has a local extremum at x_0, then $f'(x_0) = 0$.

Make sure you know what the above result is saying: the existence of a local extrema at the point x_0 coupled with differentiability in an interval surrounding x_0 implies $f'(x_0) = 0$. *The result does not say: when $f'(x_0) = 0$, the point $(x_0, f(x_0))$ is a local extrema for f.*

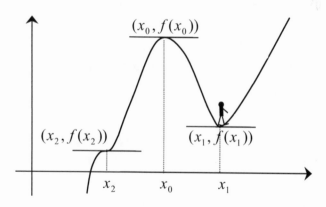

Figure 5.9: Local Extrema and Saddle Point

In **Figure 5.9**, the graph of f has a horizontal tangent line at the three points $(x_2, f(x_2))$, $(x_0, f(x_0))$ and $(x_1, f(x_1))$.

This implies that $f'(x_2) = f'(x_0) = f'(x_1) = 0$. However, f has local extrema only at $(x_0, f(x_0))$ and $(x_1, f(x_1))$. The point $(x_2, f(x_2))$ is known as a *saddle point*, which can be envisioned as a level resting spot on the side of an ascending or descending portion of a functional curve.

To prove our MAJOR result for the case of a local maximum, we appeal to the fundamental expression:

$$\frac{f(x_0 + dx) - f(x_0)}{dx} = f'(x_0)$$

Notice that $f(x_0 + dx) - f(x_0) \leq 0$ for all x values in a neighborhood of x_0, irregardless if the infinitesimal dx is positive or negative. Dividing $f(x_0 + dx) - f(x_0)$ by dx, one obtains $f'(x_0) \geq 0$ or $f'(x_0) \leq 0$ depending on the sign of dx. This immediately leads to the conclusion $f'(x_0) = 0$.

The one remaining issue is how to distinguish, without graphing, a local maximum from a local minimum, or either of the local extrema from a saddle point. But, before we address this issue, we need to have a short discussion of *increasing* and *decreasing* functions. We also need to define *critical point*.

<u>Definition</u>: a *critical point* x_0 is simply a point where $f'(x_0) = 0$.

The sign of f' is a very important indicator when analyzing functions, for it is the sign of f' that will tell us if f is *increasing* or *decreasing*. By knowing where f is *increasing* or *decreasing* with respect to a point $(x_0, f(x_0))$ with $f'(x_0) = 0$, one can determine the exact nature of the *critical point* in terms of the behavior of the function f. If $f'(x) > 0 \ \& \ dx > 0$, then $f(x + dx) - f(x) = f'(x)dx > 0$, which immediately leads to the cause-and-effect relationship $x + dx > x \Rightarrow f(x + dx) > f(x)$.

In general, if $f' > 0$ throughout an interval, we have that $f(b) > f(a)$ for any two points a, b in the interval with $b > a$: accordingly, f is said to be *increasing* on the interval. As shown by **Figure 5.10**, our now-familiar stick figure would be walking up a hill as it traverses the graph of f from left to right in an interval where the slopes f' are positive. On the other hand, if $f' < 0$ throughout an interval, we will have that $f(b) < f(a)$ for any two points a, b in the interval with $b > a$: hence, f is said to be *decreasing* on the interval. Again, by **Figure 5.10**, our figure is shown as walking down a hill in an interval where the slopes f' are negative.

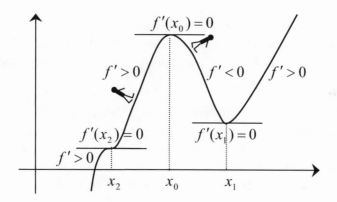

Figure 5.10: First Derivative Test

Figure 5.10 is a complete graphical layout of the First Derivative Test. The First Derivative Test, hereon abbreviated as FDT, forms the basis for a powerful two-step methodology for investigating local extrema:

$\overset{1}{\mapsto}$: *Identify critical points* by solving the equation $f'(x) = 0$.

$\overset{2}{\mapsto}$: *Characterize* the critical points found in Step 1 as to local maximum, local minimum, or saddle point.

Now we are ready to formally state the FDT.

First Derivative Test (FDT)

Precondition: let x_0 be a solution to $f'(x_0) = 0$

    1) If $x < x_0 \Rightarrow f'(x) < 0$ & $x > x_0 \Rightarrow f'(x) > 0$,

Then f has a local minimum at x_0.

    2) If $x < x_0 \Rightarrow f'(x) > 0$ & $x > x_0 \Rightarrow f'(x) < 0$,

Then f has a local maximum at x_0.

    3) If f' has the same sign on either side of x_0,

Then f has a saddle point at x_0.

In **Figure 5.10**, the point x_0 corresponds to a local maximum or local hilltop. As shown, there is an interval surrounding x_0 where one strenuously climbs up to the hilltop ($x < x_0$ and $f' > 0$), and where one carefully walks down from the hilltop ($x > x_0$ and $f' < 0$). This corresponds to 2) in the FDT as stated above. You are encouraged to correlate 1) and 3) to **Figure 5.10** where a saddle point (resting ledge) is shown at x_2 and a local minimum (valley low) is shown at x_1 .*Note: though not shown, saddle points can also occur while walking down a hill.*

    We end this subsection with four examples illustrating the use of the FDT and the associated two-step process for identifying and characterizing local extrema.

Ex 5.4.6: Find the local extrema for $f(x) = x^2 - x - 6$.

$\overset{1}{\mapsto} : f'(x) = 2x - 1 \mapsto$

$f'(x) = 0 \Rightarrow 2x - 1 = 0 \Rightarrow x = \frac{1}{2}$

    ****

$\overset{2}{\mapsto} : x < \frac{1}{2} \Rightarrow f' < 0$ & $x > \frac{1}{2} \Rightarrow f' > 0$

Therefore, by the FDT $f(x) = x^2 - x - 6$ has a local minimum at $\frac{1}{2}$. *Finally, we have* $f(\frac{1}{2}) = -\frac{25}{4}$, which completes Step 2.

Note1: when we find local extrema, we not only identify critical points, but also characterize them and evaluate the function (obtain the output values) for these same critical points. Many students think the job is over once the critical points have been identified. This is simply not true.

Note2: The result matches what would be obtained if using techniques from intermediate algebra. The quadratic function f has x intercepts at -2 and 3; hence, the local extrema lays half way in between at $\frac{1}{2}$. Since the coefficient of x^2 is positive, $f(\frac{1}{2}) = -\frac{25}{4}$ is a local minimum.

Ex 5.4.7: Find the local extrema for $f(x) = 2x^3 - 9x^2 - 24x + 1$.

$$\overset{1}{\mapsto} : f'(x) = 6x^2 - 18x - 24 = 6(x-4)(x+1) \mapsto$$
$$f'(x) = 0 \Rightarrow 6(x-4)(x+1) = 0 \Rightarrow x = 4, -1$$
****

$$\overset{2}{\mapsto} : f'(x) > 0 \Rightarrow x \in (-\infty, -1) \cup (4, \infty)$$
$$f'(x) < 0 \Rightarrow x \in (-1, 4) \mapsto$$
$$+ + + + \overset{\cap}{[-1]}_{0} - - - - \overset{\cup}{[4]}_{0} + + + +$$

The new item in Step 2 is called a *sign-change chart*, a simple but valuable graphical technique that allows us to visualize all of the information obtained in both steps. The sign-chart chart is a stylized number line where the two critical points $\overset{\cap}{[-1]}_{0}$ and $\overset{\cup}{[4]}_{0}$ are identified as such by the small zero below. Critical points are characterized as local maxima or minima by the symbols $\cap$ and $\cup$. If a saddle point were present, we would simply use the letter S. The sign of f' relative to the two critical points is indicated by the sequence $+ + + + - - - - + + + +$.

Finally, in order to complete the job, we must evaluate f at the critical points -1 and 4. We obtain $f(-1) = 14$ and $f(4) = -111$.

Ex 5.4.8: Find the local extrema for $f(x) = x^4 - 2x^2$.

Using highly condensed notation, we have:

$$\overset{1}{\mapsto}: f'(x) = 4x^3 - 4x = 4x(x-1)(x+1) \mapsto$$
$$f'(x) = 0 \Rightarrow 4x(x-1)(x+1) = 0 \Rightarrow x = 0,1,-1$$
$$****$$

$$\overset{2}{\mapsto}: f'(x) > 0 \Rightarrow x \in (-1,0) \cup (1,\infty)$$
$$f'(x) < 0 \Rightarrow x \in (-\infty,-1) \cup (0,1) \mapsto$$
$$----\overset{\cup}{[-1]}_0+++\,+\overset{\cap}{[0]}_0----\overset{\cup}{[1]}_0++++\mapsto$$
$$f(-1) = -1,\, f(0) = 0,\, f(1) = -1$$

Ex 5.4.9: Find the local extrema for $f(x) = x^3 e^{-x}$.

$$\overset{1}{\mapsto}: f'(x) = 3x^2 e^{-x} - x^3 e^{-x} = x^2(3-x)e^{-x} \mapsto$$
$$f'(x) = 0 \Rightarrow x^2(3-x)e^{-x} = 0 \Rightarrow x = 0,3$$
$$****$$

$$\overset{2}{\mapsto}: f'(x) > 0 \Rightarrow x \in (-\infty,0) \cup (0,3)$$
$$f'(x) < 0 \Rightarrow x \in (3,\infty)$$
$$++++\overset{S}{[0]}_0++++\overset{\cap}{[3]}_0----\mapsto$$
$$f(0) = 0,\, f(3) = 27e^{-3} = 1.344...$$

In this example, we encounter a saddle point at $x = 0$ while climbing up to the local maximum at $x = 3$. *Note: This is a great place for you to stop and thoroughly review the concept of continuity (Section 4.4) before continuing with the next subsection.*

5.4.4) Finding Absolute Extrema

Continuous functions have many deep and interesting properties that would be examined in detail via any *real analysis* (see note at bottom of page) course. One of these deeper properties, stated below without proof, is the essential starting point for the investigative technique developed in this subsection:

The Absolute Extrema Property

Let the function f be continuous on a closed interval $[a,b]$. Then, f has an absolute maximum and absolute minimum on $[a,b]$.

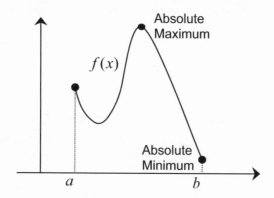

Figure 5.11: Continuity and Absolute Extrema

In **Figure 5.11**, the function f is continuous on $[a,b]$. Recalling Section 4.4, this means that one can make a smooth pencil trace—no hops, skips, or impossible leaps—from the point $(a, f(a))$ to the point $(b, f(b))$ when graphing $f(x)$. Accordingly, the path so traced will have an absolute high and an absolute low point as shown.

Note: real analysis is the formal study of limits, in particular, limits as they apply to functions where both the inputs and outputs are real numbers. Two theorems foundational to the thorough logical development of many of the deeper properties associated with functions (such as the absolute extrema property) are the Bolzano-Weierstrass and Heine-Borel theorems. Both of these theorems can be found in any standard real analysis text.

Figure 5.11 also suggests a methodology for obtaining the absolute extrema, given the function f is continuous on the closed interval $[a, b]$. As shown, the absolute maximum is also a relative maximum and the absolute minimum occurs at an endpoint. In general, absolute extrema occur either 1) inside an interval or 2) at one/both endpoints. When absolute extrema occur inside an interval, they correspond to local extrema occurring inside the same interval. Otherwise, absolute extrema occur at one or both endpoints. This suggests the following procedure for finding the absolute extrema for a continuous function f on $[a, b]$.

Let f be continuous on $[a, b]$ and differentiable on (a, b). To find the absolute extrema of f, perform the four steps below:

$\overset{1}{\mapsto}$: Find the critical points of f within $[a, b]$. Recall that critical points are points x where $f'(x) = 0$.

$\overset{2}{\mapsto}$: If need be, also find those points where f' does not exist within $[a, b]$. Such points may be few and far between. But, remember that the function f itself must be continuous—no exceptions—at these same points in order for the underlying theory to apply.

$\overset{3}{\mapsto}$: Evaluate f at all points found in step 1), in step 2), and at the two endpoints a, b.

$\overset{4}{\mapsto}$: Order the functional values obtained in 3) from largest to smallest. The largest value is the absolute maximum; the smallest is the absolute minimum.

Ex 5.4.10: Find the absolute extrema for the function $f(x) = x^3 - 3x^2 - 24x + 2$ on the closed interval $[-6, 5]$.

To start, we notice that the function f is continuous on $[-6, 5]$ since it is a well-behaved polynomial. Polynomials do not have breaks or *hiccups* in their graphs—anywhere! Hence, we are guaranteed that f has absolute extrema on $[-6, 5]$ allowing us to continue with our quest.

$\overset{1}{\mapsto}: f'(x) = 3x^2 - 6x - 24 = 3(x-4)(x+2) \mapsto$
$f'(x) = 0 \Rightarrow 3(x-4)(x+2) = 0 \Rightarrow x = 4,-2 \mapsto$
$4 \in [-6,5] \& -2 \in [-6,5]$

****

$\overset{2}{\mapsto}:$ Skip since f' exists everywhere on $(-\infty, \infty)$

****

$\overset{3}{\mapsto}:$ By the subsection discussion, $x = -6,-2,4,5$ are the only possible points where f can have an absolute extrema. Evaluating f at each of the above x values, we have:
$$f(-6) = -178, f(-2) = 30,$$
$$f(4) = -78, f(5) = -26$$

****

$\overset{4}{\mapsto}:$ After ordering, $f(-2) = 30$ is the absolute maximum and $f(-6) = -178$ is the absolute minimum.

Ex 5.4.11: Find the absolute extrema for the function $f(x) = x^3 - 3x^2 - 24x + 2$ on the closed interval $[0,5]$.

The only difference between this example and the previous example is the change of interval from $[-6,5]$ to $[0,5]$. In our new Step 1) below, one of the two critical points is not in the interval $[0,5]$. Hence, it is not a point for consideration in Step 3).

Note: Evaluating f at all critical points found in Step 1), even those critical points not in the interval of interest, is a common source of error when working these problems.

$\overset{1}{\mapsto}: f'(x) = 3x^2 - 6x - 24 = 3(x-4)(x+2) \mapsto$
$f'(x) = 0 \Rightarrow 3(x-4)(x+2) = 0 \Rightarrow x = 4,-2 \mapsto$
$4 \in [0,5] \& -2 \notin [0,5]$

****

$\overset{2}{\mapsto}$: Skip
****

$\overset{3}{\mapsto}$: Here, the points $x = 0,4,5$ are the only possible points where f can have an absolute extrema. Evaluating f at each of the above x values, we have:
$$f(0) = 2, f(4) = -78, f(5) = -26$$
****

$\overset{4}{\mapsto}$: $f(0) = 2$ is the absolute maximum, and $f(4) = -78$ is the absolute minimum.

Ex 5.4.12: Find the absolute extrema for the function $f(x) = (2x-1)e^{-x}$ on the closed interval $[0,5]$.

Products of simple exponential and polynomial functions are continuous. Hence the function above has absolute extrema on the interval $[0,5]$. Continuing:

$\overset{1}{\mapsto} : f'(x) = -(2x-1)e^{-x} + 2e^{-x} = (3-2x)e^{-x} \mapsto$
$f'(x) = 0 \Rightarrow 3 - 2x = 0 \Rightarrow x = 1.5 \mapsto$
$1.5 \in [0,5]$
****

$\overset{2}{\mapsto}$: Skip since f' exists everywhere on $(-\infty, \infty)$
****

$\overset{3}{\mapsto}$: The three candidate x values are $x = 0,1.5,5$. Evaluating f at the same points results in:
$$f(0) = -1, f(1.5) = .44626, f(5) = .0606.$$
****

$\overset{4}{\mapsto}$: The absolute maximum is $f(1.5) = .44626$, and the absolute minimum is $f(0) = -1$.

Ex 5.4.13: Find the absolute extrema for the function $f(x) = \dfrac{x^2}{\ln(x)-1}$ on the closed interval $[4,6]$.

A little preliminary discussion is in order. The function f is not defined at the two x values $0\ \&\ e$. Subsequently, the graph of f does not exist at either 0 or e, implying 1) there is a hole in the graph at these two x values and 2) f is discontinuous at the same. Now, if we would have been asked to find the absolute extrema of f on $[0,4]$, we would have had to capitulate. This is because f has two *bad points* $0\ \&\ e$ inside the interval $[0,4]$, which implies functional discontinuity on $[0,4]$. *But none of this is true for the interval* $[4,6]$, *our interval of interest!* Thus, the theory applies, and one can proceed as before.

$$\overset{1}{\mapsto}: f'(x) = \frac{(\ln(x)-1)2x - \frac{1}{x}(x^2)}{(\ln(x)-1)^2} = \frac{x(2\ln(x)-3)}{(\ln(x)-1)^2} \mapsto$$

$$f'(x) = 0 \Rightarrow 2\ln(x) - 3 = 0 \Rightarrow$$

$$\ln(x) = 1.5 \Rightarrow x = e^{1.5} \mapsto$$

$$e^{1.5} = 4.48 \in [4,6]$$

****

$\overset{2}{\mapsto}$: Notice that f' is also undefined at 0 or e. However, we shall disregard 0 or e since again, these points lay outside the interval $[4,6]$ of interest. *Note: It is really immaterial to even discuss either 0 or e in the context of f' since the function f itself is not even defined at 0 or e.*

****

$\overset{3}{\mapsto}$: The three candidate x values are $x = 4$, $x = e^{1.5} = 4.48$, and $x = 6$. Evaluating f at the same:

$$f(4) = 41.42,\ f(4.48) = 40.7,\ f(6) = 45.46.$$

****

127

$\overset{4}{\mapsto}$: Finally, the absolute maximum is $f(6) = 45.46$ and the absolute minimum is $f(4.48) = 40.7$.

Ex 5.4.14: Find the absolute extrema for the function $f(x) = x^{\frac{2}{3}}$ on the closed interval $[-1,1]$.

The function itself is continuous on $[-1,1]$ (in fact, on $(-\infty,\infty)$ since we can take the cube root of any number and subsequently square it (the true meaning of $x^{\frac{2}{3}}$).

$$\overset{1}{\mapsto}: f'(x) = \frac{2}{3}x^{\frac{-1}{3}} = \frac{2}{3\sqrt[3]{x}}$$

For this function, there is no x value where $f'(x) = 0$.
****

$\overset{2}{\mapsto}$: Notice that even though f is defined at $x = 0$, its derivative f' fails to be defined at the same. Since $x = 0$ is inside the interval $[-1,1]$, we will need to investigate this curious phenomena just like we would investigate a normal critical point where $f'(x) = 0$. A simple graph (**Figure 5.12**) will help.

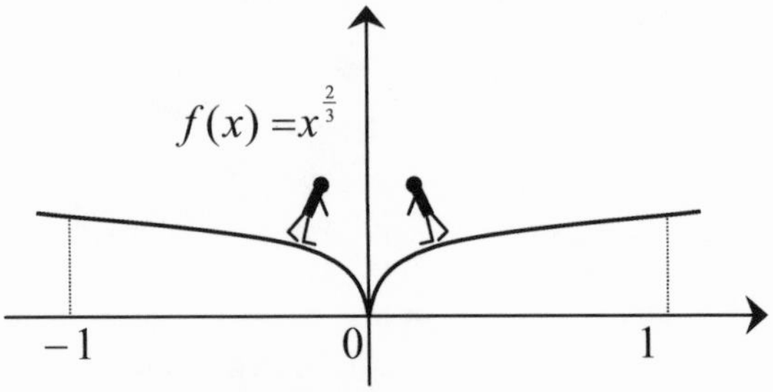

Figure 5.12: Graph of $f(x) = x^{\frac{2}{3}}$

As we move down into the low point at $x = 0$, the two converging and merging walls become very, very steep. This leads to an exact vertical slope at $x = 0$. Vertical slopes have *rise with no run*. The *no run* situation creates a division by zero in the slope formula $m = \dfrac{rise}{run}$ and will render $f'(x)$ as undefined at the point in question. In this case it is $x = 0$. In common terms, the graph of the function f has an extremely sharp knife edge at the point $x = 0$. Subsequently, in the absence of any sort of rounding at $x = 0$, the function f will not support a slope at the same. All such points, when they occur inside an interval of interest, must be investigated along with regular critical points. As this example shows, the point may be exactly the value we seek.

****

$\overset{3}{\mapsto}:$ The three candidate x values are $x = -1$, $x = 0$ and $x = 1$. Evaluating f at the same:
$$f(-1) = 1, f(0) = 0, f(1) = 1.$$

****

$\overset{4}{\mapsto}:$ The absolute maximum is $f(1) = f(-1) = 1$ and the absolute minimum is $f(0) = 0$.

Note: it is quite alright for two points to share the honors regarding absolute extrema, or local extrema for that matter.

In light of the last example, we close this subsection with an expanded definition of critical point.

<u>Expanded Definition of Critical Point</u>: a *critical point* x_0 for a function f is simply a point *in the domain of* f where one of the following happens: 1) $f'(x_0) = 0$ or 2) $f'(x_0)$ is not defined.

5.4.5) Geometric Optimization

     This, the last subsection, addresses those infamous word problems as they apply to the optimization of geometric quantities. We will not give you 'six easy steps' as do many authors. At least in my personal experience, the way to mastery of word problems is by chewing on, failing at, and, finally, succeeding at the task. Lots of struggling practice gives one a feel for what works and what doesn't. Don't despair when first starting out towards the end goal of mastery, for mastery is reached by a road paved with hours of practice.

Note: both in quality circles and in this book, the word optimize can mean one of three things: maximize, minimize, or nominalize.

     Two major points must be made before proceeding on with three classic optimization examples. The first major point is that the drawing of a diagram or picture representing the situation at hand, no matter how silly or simple, is a great aid in solving word problems. Diagrams allow for the visualization of non-linear and/or spatial relationships, enabling better right-left brain integration, whereas equations by themselves are primarily linear or left brain in nature. The other major point is that when an equation is finally put together that algebraically represents the quantity to be optimized; it must be expressed in terms of a differentiable function that accurately models the situation when restricted to a suitable domain. Once such a function is constructed, the hard work is done. We can then go ahead and proceed as we did in the two previous subsections.

Ex 5.4.15: <u>The Famous Girder Problem</u>: Two people at a construction site are rolling steel girders down a corridor 8 feet wide into a second corridor $5\sqrt{5}$ feet wide and perpendicular to the first corridor. What is the length of the longest girder that can be rolled from the first corridor into the second corridor and continued on its journey in the construction site? Assume the girder is of negligible thickness.

Note1: The famous girder problem started to appear in calculus texts circa 1900. My father first encountered the girder problem in 1930 when he was still an engineering student at Purdue. Thirty-six years later, I too encountered and defeated it after eight hours of continuous struggle. The girder problem still appears in modern calculus textbooks disguised and somewhat watered down as a geometric optimization problem. The problem is famous because of the way it thoroughly integrates the principles of plane geometry, algebra, and differential calculus. My experience as a teacher has been—when assigning this gem—that 'many try, but few succeed.' In this book, we will guide you through the entire thought process needed to obtain a solution. Your job is to thoroughly understand the thought process as presented.

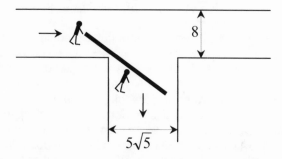

Figure 5.13: Schematic for Girder Problem

Figure 5.13 is a suggested schematic for the girder problem where all the information given in the verbal description is visually laid out in a spatial format (remember my opening comments on left-right brain integration). Two things are readily apparent: 1) the girder as pictured will roll around the corner without jamming, allowing the two workers to continue their task; and 2) there are longer girders that will jam when the two workers attempt to roll them around the corner.

Girder Problem Objective

The objective of the girder problem is to find the
longest possible girder that will roll around the corner.

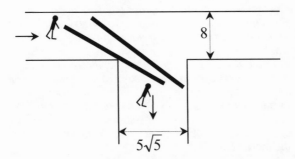

Figure 5.14: Use of Pivot Point in Girder Problem

Continuing, **Figure 5.14** depicts a thought experiment that shows the most geometrically advantageous way of rolling girders around the corner. Two girders of equal length are pictured in transit from the first hallway to the second. The rightmost girder has an imaginary pivot point near the middle of the entrance to the second hallway. As shown, this girder will probably jam and a backup/start-over will be necessary. On the other hand, if the workers use the left-most entrance wall as a pivot point, the girder will definitely roll around the corner. We conclude that the longest possible girder that can be rolled around the corner must use the left-most entrance wall as a pivot point.

Let's take the thought experiment to an additional level of complexity. Imagine that our pivoting girder has spring-loaded extenders on both ends. These extenders extend or compress in order to maintain contact with the two walls as shown in **Figure 5.15** while the girder rotates around the pivot point. During the rotation process, the length $\overline{AB}$ from extender tip to extender tip will vary. Just going into the turn, the rightmost extender is shorter than the length of the girder itself; but the leftmost extension could be several times the length of the girder. The opposite situation applies when coming out of the turn. Moving through the turn, as long as $\overline{AB}$ is contracting while the girder is rotating, then a fixed beam of length $\overline{AB}$ will not make it around the corner. Once $\overline{AB}$ stops contracting, this shortest $\overline{AB}$ corresponds to the length of the longest rigid girder that can be rotated around the corner.

Note2: The original problem was stated in terms of maximizing a quantity. The problem has turned out to be one of minimizing a quantity, in particular the geometric quantity $\overline{AB}$.

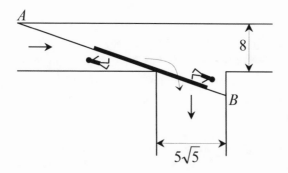

Figure 5.15: Girder Extenders

At this point, we are ready to construct a bare-bones geometric diagram (**Figure 5.16**), which will guide us as we begin to abstract our physical situation into a mathematical model (i.e. start the necessary function formulation).

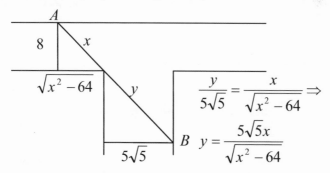

Figure 5.16: Geometric Abstraction of Girder Problem

Algebraically, the problem is to minimize the diagonal length $\overline{AB}$ where $\overline{AB} = x + y = x + \dfrac{5\sqrt{5}x}{\sqrt{x^2 - 64}}$.

Define a length function L as follows:

$$L(x) = x + \frac{5\sqrt{5}x}{\sqrt{x^2 - 64}}$$

Even though L is algebraically defined on the domain $(-\infty, -8) \cup (8, \infty)$, L only makes physical sense (in the context of our problem) on the interval $(8, \infty)$ where the output values are always positive. Notice that the output values for L grow without bound as the domain values x approach 8 from the right. The output values also grow without bound with increasing x values on the left. Both end-of-interval situations match the physical context. Again, the sweet spot will be that point where L is minimized. **Figure 5.17** is a notational diagram of how we have intuited the behavior of the function L thus far.

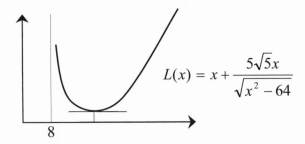

Figure 5.17: Notional Graph of $L(x)$

After four pages of preliminary discussion and analysis, we are ready to analyze the function $L(x)$ using the two steps in Subsection 5.4.3. This is the way it always goes with word problems: lots of preliminary consideration and conceptualization before the function is finally framed. In this key example, we have allowed you to peak at this *not-so-cut-and-tried* process first hand.

$$\overset{1}{\mapsto} : L'(x) = 1 + 5\sqrt{5}\left[\frac{(x^2-64)^{\frac{1}{2}}\cdot 1 - \frac{1}{2}(x^2-64)^{-\frac{1}{2}}2x\cdot x}{\{(x^2-64)^{\frac{1}{2}}\}^2}\right] \Rightarrow$$

$$L'(x) = 1 - \frac{320\sqrt{5}}{(x^2-64)^{\frac{3}{2}}} = \frac{(x^2-64)^{\frac{3}{2}} - 320\sqrt{5}}{(x^2-64)^{\frac{3}{2}}} \mapsto$$

$$L'(x) = 0 \Rightarrow (x^2-64)^{\frac{3}{2}} - 320\sqrt{5} = 0 \Rightarrow$$

$$x^2 - 64 = (320\sqrt{5})^{\frac{2}{3}} \Rightarrow x^2 - 64 = 80 \Rightarrow$$

$$x = 12\,ft$$

We also have that $L(12) = 12 + \dfrac{5\sqrt{5}(12)}{\sqrt{12^2 - 64}} = 27\,ft$.

****

$\overset{2}{\mapsto} :$ Here Step 2 is used to confirm our conceptual analysis.

$$L'(x) < 0 \Rightarrow x \in (8,12)$$
$$L'(x) > 0 \Rightarrow x \in (12,\infty) \mapsto$$

$$\overset{\cup}{(8----[12]+++++++...\infty)}_{0}$$

EUREKA!!!—Remember Archimedes?

The famous girder problem has finally come to a close and the answer is 27 feet. This is the length of the longest girder that can be moved around the corner without destroying a wall.

Our next two examples are not nearly as involved. Therefore, many of the conceptualization and setup details will be left to the reader as part of the learning process. But don't forget, we have just presented, in excruciating detail, one of the hardest word problems in elementary calculus. Anything else is cake.

Ex 5.4.16: A small rectangular box with no top is to be made from a square piece of cardboard that measures 10 inches on a side. In order to make the box, a small square is first cut out from each of the four corners. Next, the remaining part of the side is folded up as shown in **Figure 5.18**. What size square must be cut out to give a box of maximum volume? What is the maximum volume?

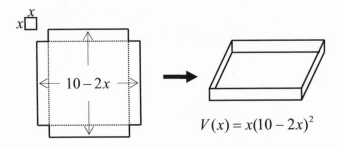

$$V(x) = x(10 - 2x)^2$$

Figure 5.18: Box Problem

The function describing the volume of the box so made is given by $V(x) = x(10 - 2x)^2$, which is algebraically defined for all real numbers. Hence, the algebraic domain is given by $(-\infty, \infty)$. However, in the context of our real-world problem, the physical domain is $[0,5]$. The endpoint $x = 0$ corresponds to no cut at all, and the endpoint $x = 5$ corresponds to the whole piece being quartered. Each of these extreme conditions leads to zero volume for the folded-up box. For all x values in $(0,5)$, it is intuitively obvious that we will get a positive volume. Since V is continuous on $[0,5]$, we know that V must achieve a maximum. Also, since $V(0) = V(5) = 0$, we know that the maximum must occur within the interval $[0,5]$ and at a domain value where $V' = 0$. We continue our analysis of V using absolute extrema methods.

$$\overset{1}{\mapsto}: V'(x) = x \cdot 2(10 - 2x)^1(-2) + 1 \cdot (10 - 2x)^2 \Rightarrow$$
$$V'(x) = 4(5 - x)(5 - 3x) \mapsto$$
$$V'(x) = 0 \Rightarrow x = 5, \tfrac{5}{3} \mapsto$$
$$\tfrac{5}{3} \in (0,5)$$

Note: the critical point $x = 5$ is not included since it already has gained admittance to the examination process by being one of the two endpoints.
****

$\overset{2}{\mapsto}:$ Since V' is defined everywhere, this step does not apply.
****

$\overset{3}{\mapsto}:$ The three candidate x values are $x = 0, x = \tfrac{5}{3}$, and $x = 5$. Evaluating V at the same:
$$V(0) = 0, V(\tfrac{5}{3}) = \tfrac{2000}{27} = 74.07 in^3, V(5) = 0.$$
****

Now we can verbally answer the two original questions. Our small squares should be $1\tfrac{2}{3}$ inches to a side. Finally, the volume of a box constructed to this specification—the largest possible using the given configuration and resources—is given by $74.07 in^3$.

Ex 5.4.17: The strength of a beam with rectangular cross-section is directly proportional to the product of the width and the square of the depth (thickness from top to bottom). Find the shape of the strongest beam that can be cut from a cylindrical log of diameter d feet as shown in **Figure 5.19**.

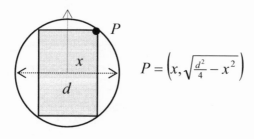

Figure 5.19: Beam Problem

Let P be a corner point as shown. Then the strength $S = S(x)$ is given by the expression $S(x) = k(2x)\left(2\sqrt{\frac{d^2}{4} - x^2}\right)^2$ where the letter k is the constant of proportionality. The function S is physically defined on the interval $[0, \frac{d}{2}]$ and can be immediately reduced to $S(x) = 2kx(d^2 - 4x^2)$. Also, observe the endpoint equality $S(0) = S(\frac{d}{2}) = 0$, which means beams cut to the extremes of one dimension have little or no remaining strength.

Note: The reduced function S has $(-\infty, \infty)$ as its algebraic domain. Hopefully by now you are beginning to see the difference between the algebraic and physical domains when solving word problems.

$\overset{1}{\mapsto} : S'(x) = 2k(d^2 - 12x^2) \mapsto$

$S'(x) = 0 \Rightarrow x = \frac{d}{2\sqrt{3}}, -\frac{d}{2\sqrt{3}} \mapsto$

$\frac{d}{2\sqrt{3}} \in (0, \frac{d}{2})$

****

$\overset{2}{\mapsto} :$ This step does not apply.

****

$\overset{3}{\mapsto} :$ The only candidate value left after endpoint considerations is $x = \frac{d}{2\sqrt{3}}$. Evaluating S, we have

$$S(\tfrac{d}{2\sqrt{3}}) = 2k(\tfrac{d}{2\sqrt{3}})(d^2 - [\tfrac{d}{2\sqrt{3}}]^2) = \frac{11kd^3}{12\sqrt{3}}.$$

****

Summary: in order to maximize strength, our beam should be cut to a total width of $\frac{d}{\sqrt{3}}$ and total depth of $\frac{2\sqrt{6}}{3}d$. The strength of a beam constructed to this specification is $\dfrac{11kd^3}{12\sqrt{3}}$.

Always Remember

No word problem is complete until all the original questions, as initially stated in the problem, are answered.

Our next example demonstrates the Pythagorean Theorem using the optimization methods of calculus. As does the girder problem, **Ex 5.4.15**, this example unites several geometric principles in order to complete the demonstration of the Pythagorean Theorem. Unlike the girder problem, the use of calculus is optional, hence the use of the word *demonstrates* instead of *proves*.

Note: The reader is challenged in the Section Exercise # 9 to discover where the use of calculus is optional.

Before we embark on our calculus-based demonstration, we will present one traditional algebraic proof of the Pythagorean Theorem attributed to the Indian mathematician Bhaskara.

<u>Statement of the Pythagorean Theorem</u>: Suppose we have a right triangle where C is the length of the hypotenuse (side opposite the right angle). Let A and B be the lengths of the remaining two sides. Then

$$A^2 + B^2 = C^2$$

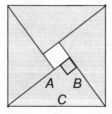

Figure 5.20: Bhaskara's Diagram

139

<u>Proof</u>: Bhaskara's stroke of genius was the arranging of the original right triangle and three identical replicates into the large square as shown in **Figure 5.20**.

$\mapsto^{1}$: From **Figure 5.20**

$$AREA_{bigsquare} = AREA_{littlesquare} + 4 \cdot (AREA_{onetriangle})$$

$\mapsto^{2}$: Substitute the values A, B, C into the above equality
In order to complete the proof

$$C^2 = (A - B)^2 + 4(\tfrac{1}{2} AB) \Rightarrow$$
$$C^2 = A^2 - 2AB + B^2 + 2AB \Rightarrow$$
$$C^2 = A^2 + B^2 \Rightarrow A^2 + B^2 = C^2 \;\therefore$$

Ex 5.4.18: <u>Demonstration of the Pythagorean Theorem</u>: **Figure 5.21** is the geometric focus for our demonstration. The figure is named '**Carolyn's Cauliflower**', in honor of my wife Carolyn who suggested that the structure resembled a head of cauliflower.

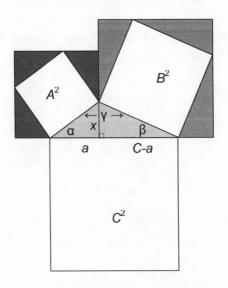

Figure 5.21: Carolyn's Cauliflower

The objective of our demonstration will be to show that the three squares A^2, B^2, C^2 constructed from the lengths of the three sides of the light-gray triangle shown in **Figure 5.21** satisfy $A^2 + B^2 = C^2$ when the three interior angles α, β, γ of the same triangle satisfy the equality $\alpha + \beta = \gamma$. Since $\alpha + \beta + \gamma = 180^0$ for any triangle, the equality $\alpha + \beta = \gamma$ is equivalent to the condition $\gamma = 90^0$, which in turn implies the light-gray triangle [which we will denote by (α, β, γ) throughout the remainder of the demonstration] is a right triangle.

From the stated objective, a careful reader will observe that in reality, we are about to demonstrate the converse of the Pythagorean Theorem:

$$A^2 + B^2 = C^2 \Rightarrow \alpha + \beta = \gamma.$$

However, since all critical steps in the subsequent demonstration are reversible in their implications, our final result can be stated

$$A^2 + B^2 = C^2 \Leftrightarrow \alpha + \beta = \gamma,$$

which succinctly combines the two conclusions associated with the Pythagorean Theorem and the Pythagorean Converse.

To start our demonstration, let $C > 0$ be the fixed length of an arbitrary line segment placed in a horizontal position. Let a be a fixed but arbitrary point on the open interval $(0, C)$ which cuts the line segment into two sub-lengths: a and $C - a$. Let $x \geq 0$ be the *adjustable length* of a perpendicular line segment erected at point a. As such, x serves as the altitude for the *adjustable triangle* (α, β, γ) as shown in **Carolyn's Cauliflower**. Initially, C serves as the hypotenuse for (α, β, γ) when x is small but obviously relinquishes this role when x is sufficiently large. *Thus, we will restrict x values to those values where C still serves as a legitimate hypotenuse (See Section Exercise 10).*

The sum of the two square areas A^2 and B^2 in **Figure 5.21** can be determined in terms of a and x as follows:

$$A^2 + B^2 = (a+x)^2 + ([C-a]+x)^2 - 2Cx.$$

The terms $(a+x)^2$ and $([C-a]+x)^2$ are the areas of the left and right outer enclosing squares, and the term $2Cx$ is the combined area of the eight shaded triangles expressed as an equivalent rectangle. Next, we define an *area-excess function*

$$E(x) = \{A^2 + B^2 - C^2\}^2 \geq 0,$$

which is the square of the difference between $A^2 + B^2$ and the hypotenuse C^2. Substituting for $A^2 + B^2$ in $E(x)$, we have

$$E(x) = \{(a+x)^2 + ([C-a]+x)^2 - 2Cx - C^2\}^2 \Rightarrow$$
$$E(x) = \{2a^2 + 2x^2 - 2Ca\}^2 \Rightarrow$$
$$E(x) = 4\{a[C-a] - x^2\}^2$$

In terms of the function $E(x)$, our objective is to use extrema methods to find the geometric conditions for which $E(x) = 0$.
In turn, $E(x) = 0 \Leftrightarrow A^2 + B^2 - C^2 = 0 \Leftrightarrow A^2 + B^2 = C^2$.

$\overset{1}{\mapsto} : E'(x) = -16y\{a(C-a) - x^2\} \mapsto$
$E'(x) = 0 \Rightarrow x = 0 \,\&$
$a(C-a) - x^2 = 0 \Rightarrow x = \sqrt{a(C-a)}$
****

$\overset{2}{\mapsto} :$ Does not apply since $0 < a < C \Rightarrow a(C-a) > 0$
****

$\mapsto$: The first of the two candidate critical points is $x = 0$. We can immediately discount this case since it leads to a collapsed triangle as shown in **Figure 5.22**. We also have that $E(0) = 4\{a[C-a]\}^2 > 0$, which implies $A^2 + B^2 \neq C^2$, more precisely $A^2 + B^2 < C^2$.

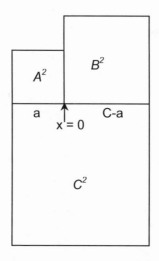

Figure 5.22: Collapsed Triangle

To examine the second critical point—actually an entire locus of critical points x_{cp} defined by the equation $a[C-a] - x_{cp}^{2} = 0$—we first substitute $x_{cp} = \sqrt{a(C-a)}$ into $E(x)$ to obtain

$$E(x_{cp}) = E(\sqrt{a(C-a)}) \Rightarrow$$
$$E(x_{cp}) = 4\{a[C-a] - \{\sqrt{a(C-a)}\}^2\}^2 .$$
$$E(x_{cp}) = 4\{a[C-a] - a(C-a)\}^2 = 0$$

Thus, the conclusion of the Pythagorean Theorem is satisfied in that

$$E(x_{cp}) = 4\{a[C-a] - a(C-a)\}^2 = 0 \Rightarrow$$
$$E(x_{cp}) = (A^2 + B^2 - C^2)^2 = 0 \Rightarrow$$
$$A^2 + B^2 = C^2$$

However, what are the geometric characteristics of an arbitrary triangle (α, β, γ) with $a[C-a] - x_{cp}^2 = 0$ constructed as shown in **Figure 5.21**? To answer this question, we first rewrite $a[C-a] - x_{cp}^2 = 0$ as a proportional equality

$$\frac{a}{x_{cp}} = \frac{x_{cp}}{C-a}.$$

Next, we examine the proportional equality in light of **Figure 5.23** where we see that this same equality establishes direct proportionality of non-hypotenuse sides for the two triangles ΔLPM and ΔMPN .

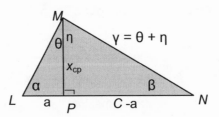

Figure 5.23: Geometry of Critical Points

Additionally, we see that both triangles have interior right angles, establishing that $\Delta LPM \approx \Delta MPN$. Thus $\alpha = \eta$ and $\theta = \beta$. Since the sum of the remaining two angles in a right triangle is 90^0 both $\alpha + \theta = 90^0$ and $\eta + \beta = 90^0$.

Combining $\eta + \beta = 90^0$ with the equality $\theta = \beta$ and the definition for γ immediately leads to $\gamma = \eta + \theta = 90^0$. This establishes the key fact that ΔLMN is a right triangle. In turn, we have the equivalency of the three conditions

$$A^2 + B^2 = C^2 \Leftrightarrow a[C-a] - x_{cp}^{\;2} = 0 \Leftrightarrow \alpha + \beta = \gamma$$

where the middle condition was obtained using optimization methods. Thus completes the calculus-based demonstration of the Pythagorean Theorem and the Pythagorean Converse.

Ex 5.4.19: Our final example in Section 5.4 comes from heat transfer and addresses a concept called the Critical Radius of Insulation. Consider a non-insulated and thin-walled hot water pipe made of copper. The heat loss q per unit length is given in macro terms as $q = h2\pi r_i(T_W - T_\infty)$ where T_W is the water temperature, T_∞ is the average temperature outside the copper pipe that acts to cool the pipe and contents, r_i is the inside radius of the thin-walled pipe—which essentially doubles as the outside radius of the pipe under the thin-walled assumption—and h is the heat-transfer coefficient (a very complex thermo-physical quantity that depends on flow and surface conditions). **Figure 5.24** shows the non-insulated hot water pipe.

Note: Albert Einstein, the father of modern Relativity Theory, abandoned the study of turbulent-flow heat transfer due to the complexity of the mathematics involved.

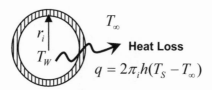

Figure 5.24: Non-insulated Hot Water Pipe

It is beyond the scope of this book to develop the expression for heat loss *per unit length of pipe* as a function of insulation thickness as measured from the center of the hot-water pipe (The interested reader is invited to research the reference in the note below). We will simply state the result:

$$q(r) = \frac{\Delta T}{\dfrac{1}{2\pi k}\ln\left[\dfrac{r}{r_i}\right] + \dfrac{1}{2\pi r h}} \quad \text{where } r \geq r_i.$$

In the above expression, r_i is the inner radius of the hot water pipe; r is the outer radius of the insulation; k is the thermal conductivity of the insulating material; h is the outside heat-transfer coefficient; and $\Delta T = T_W - T_\infty$. **Figure 5.25** depicts the now-insulated pipe.

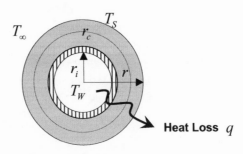

Figure 5.25: Insulated Hot Water Pipe

Note: This problem was one of the very first 'real-world' applications of calculus that I encountered as a practicing engineer. It was truly a magnificent eye-opener! For additional eye-opening calculus examples in the mechanical engineering discipline of heat transfer, the reader is referred to: Myers, Glenn; Analytical Methods in Conduction Heat Transfer 2nd Edition; Amcht Publishers; 1998.

If we let

$$R(r) = \frac{1}{2\pi k} \ln\left[\frac{r}{r_i}\right] + \frac{1}{2\pi r h},$$

then we can write

$$q(r) = \frac{\Delta T}{R(r)}.$$

The last expression states that the heat loss is equal to the temperature difference (a thermal potential) divided by the total *thermal resistance* $R(r)$ between the two extreme temperatures that comprises ΔT. Much of elementary conduction heat transfer concerns the generation of expressions for thermal resistance, which primarily depend on geometric and material characteristics. From these expressions, the various heat exchanges—or transfers—within a given thermodynamic system can be ascertained.

Our primary objective in this example is to examine the behavior of

$$R(r) = \frac{1}{2\pi k} \ln\left[\frac{r}{r_i}\right] + \frac{1}{2\pi r h} \quad \text{on the interval } [r_i, \infty).$$

Indeed $R(r)$ increases without bound as r increases without bound implying we can make the heat loss q as low as we want as long as we comply with other constraints imposed by economics or overall geometries. At the endpoint r_i we have $R(r_i) = 1/2\pi r_i$ and $q(r_i) = 2\pi r_i \Delta T$, the heat loss per unit length of non-insulated pipe. But does R initially reduce in size from its endpoint value $R(r_i) = 1/2\pi r_i$ before it starts to increase without bound?

To answer, we use the extrema methods of calculus to examine the behavior of $R(r)$ on $[r_i,\infty)$.

$$\overset{1}{\mapsto}: R'(r) = \left\{\frac{1}{r} - \frac{k}{hr^2}\right\} \mapsto$$

$$R'(r) = 0 \Rightarrow \left\{\frac{1}{r} - \frac{k}{hr^2}\right\} = 0 \Rightarrow$$

$$r = r_c = \frac{k}{h}$$

****

$$\overset{2}{\mapsto}: \quad \text{Does not apply}$$

****

$$\overset{3}{\mapsto}: R''(r) = \left\{\frac{-1}{r^2} + \frac{2k}{hr^3}\right\} \mapsto$$

$$R''(r_c) = R''\left(\tfrac{k}{h}\right) = \frac{h^2}{k^2} > 0$$

We conclude that $R(r)$ has an absolute minimum on the interval $[r_i,\infty)$ denoted by $r_c = k/h$ (see r_c in **Figure 5.25**). A proper engineering interpretation of the quantity r_c (Critical Radius of Insulation) in the context of the problem is absolutely necessary. We have two cases.

Case1: If k and h are such that $r_i < r_c = \dfrac{k}{h} < \infty$, then $R(r)$ will decrease in value on the subinterval $[r_i,r_c)$, achieve an absolute minimum at $r = r_c$, and increase without bound on the subinterval (r_c,∞).

Correspondingly, we have the heat loss $q(r)$ actually increasing $q(r) > q(r_i)$ on the interval $[r_i, r_c)$. This is because the effect of increasing the overall pipe exposure area (a direct function of radius) to T_∞ overshadows the actual thickness $(r - r_i)$ of the insulating layer even though the layer may have a reasonably low thermal conductivity k. The name 'critical radius' where $r_i < r < r_c$ implies we are doing more harm than good by applying just a $r's$ worth of insulation. This is still true for $r > r_c$, but the r value eventually will be such that $q(r) < q(r_i)$ since $R(r)$ increases without bound. Knowledge of the critical radius r_c is a prime consideration when protecting a hot-water pipe from heat loss. As one can see, it is totally dependent on the thermo-physical quantities k and h.

Case2: If k and h are such that $\dfrac{k}{h} = r_c < r_i$, then any amount of applied insulation will increase $R(r)$ from its hypothetical low value at $r = r_c$, now inside the water pipe itself.

$$\int_a^b \overset{\bullet\bullet}{\cup} dx$$

Section Exercises

1. Consider the function $f(x) = 5 + 3x - 2x^2$
    a. Find the equation of the tangent line at $x = 2$.
    b. Find the equation of the normal line at $x = 2$.
    c. Find the local extrema for f.
    d. Find the absolute extrema for f on $[-2,2]$.
    e. Find the absolute extrema for f on $[1,3]$.
    f. Find the shortest distance from the point $(2,1)$ to the graph of the function f.

2. The graphs of the two quadratic functions $f(x) = x^2 + 2$ and $g(x) = 4x - x^2$ touch each other at one common point of tangency. Find the equation of the common tangent line at this same point.

3. Consider the function $f(x) = x^3 + 2x^2 + x + 1$
      a. Find the local extrema for f.
      b. Find the absolute extrema for f on $[-\frac{1}{2}, 1]$.
      c. Use Newton's Method to evaluate the one x intercept accurate to three decimal places.

4. Concrete is being used to pour a huge slab of dimensions $3\,ft \times 150\,ft \times 150\,ft$. The thickness is correct; but, three hours into the pouring process, the two lateral dimensions are re-measured at $152\,ft$. Using linear approximation methods, quickly determine the additional amount of concrete (in cubic yards) needed at the pouring site.

5. Use linear approximation to evaluate $\sqrt{76}$ and $\sqrt[3]{29}$.

6. Find the local extrema for $f(x) = \frac{1}{4}x^4 - x^3 + 4x - 1$.

7. Consider a triangle inscribed within a unit half circle as shown below.

      a. Find the area of the largest triangle that can be inscribed in this fashion.
      b. Find the perimeter for the same.

8. Find the absolute extrema on the interval $[0,10]$ for each of the following functions: $f(x) = xe^{-x^2}$ and $g(x) = \ln(x^2 - x + 1)$.

9. Explain why the use of calculus *is optional* as a mathematical tool to locate the absolute minimum for the area-excess function $E(x)$ defined in **Ex 5.4.18**.

10. Given $C > 0$ and $0 < a < C$ as defined in **Ex 5.4.18**, find the smallest value of $x > 0$ where C no longer serves as the hypotenuse as constructed in **Figure 5.21**. Show that this value (call x_{hyp}) satisfies $0 < x_{cp} < x_{hyp}$ where $x_{cp} = \sqrt{a(C - a)}$.

11. In light of **Ex 5.4.19**, would there ever be a potential engineering situation where letting the hot-water pipe 'go bare' is a viable solution for the prevention of heat loss?

Love Triangle

Consider old Pythagoras,
A Greek of long ago,
And all that he did give to us,
Three sides whose squares now show

In houses, fields and highways straight;
In buildings standing tall;
In mighty planes that leave the gate;
And, micro systems small.

Yes, all because he got it right
When angles equal ninety—
One geek (BC), his plain delight—
One world changed aplenty!

January 2002

5.5) Process Adaptation: Implicit Differentiation

The previous section (the longest in the book) explores several elementary applications of the derivative. In this section, we are going to modify the differentiation process itself in order to find derivatives for those functions that are defined *implicitly*.

To start the discussion, consider $y = f(x) = \dfrac{1}{x^2}$. This function is defined explicitly in that we can easily see the exact algebraic expression for the input-to-output processing rule, namely $\dfrac{1}{x^2}$. Differentiation of such explicitly defined functions is easy. All we have to do is follow the rules in Section 5.3. Applying the basic power rule to $y = f(x) = x^{-2}$, we obtain

$$y = x^{-2} \Rightarrow$$
$$[y]' = [x^{-2}]' \Rightarrow$$
$$y' = -2x^{-3} = \frac{-2}{x^3}$$

You might ask, why go to such detail? It seems to be algebraic overkill. The answer is that it illustrates what is really going on in the now pretty-much automated differentiation process. In words, since the output expression for y is equal to $\dfrac{1}{x^2}$, then the derivative for y will be equal to the derivative of $\dfrac{1}{x^2}$. Thinking back to the rules of equality first learned when solving elementary linear equations, one could say that we have just added yet another rule:

Derivative Law of Equality: If $A = B$**, then** $A' = B'$

Where all equality rules can be reduced to the single common principle:

Whatever is done to one side of an equality statement must
be done to the other side of the equality statement in order to
preserve the equality statement after completion.

So, why are we having all this fuss about equality statements?
Answer: a proper understanding of equality statements is one of
the necessary preparations when learning how to differentiate
implicit functions.

It doesn't take a whole lot of algebraic manipulation to
turn $y = \dfrac{1}{x^2}$ into an *implicit function*. By simple subtraction of
equals, the function $y = \dfrac{1}{x^2}$ becomes $y - \dfrac{1}{x^2} = 0$. The expression
$y - \dfrac{1}{x^2} = 0$ is called an *implicit formulation* of y as a function of x
because the expression hints or implies that y indeed is a function
of x, but does not specifically showcase the exact algebraic and/or
transcendental formula for y in terms of x as does the explicit
formulation $y = \dfrac{1}{x^2}$. The differentiation of the implicit formulation
of $y = f(x)$ is quite easy if we use the equality rules just
presented.

$$\left[y - \frac{1}{x^2} \right]' = [0]' \Rightarrow$$

$$[y]' - \left[\frac{1}{x^2} \right]' = 0 \Rightarrow$$

$$y' + \frac{2}{x^3} = 0 \Rightarrow y' = \frac{-2}{x^3}$$

Notice, that in the above, we simply use derivative rules to
differentiate both sides of the expression, being careful to preserve
the equality during all steps in the process.

The last step is the actual solving for the derivative y' with the final answer matching (as we would expect) the result obtained from the explicit differentiation.

Suppose we create a different implicit functional form from $y = \dfrac{1}{x^2}$, say the expression $yx^2 = 1$. Could we use this as a new starting point to obtain the known final result $y' = \dfrac{-2}{x^3}$? The answer is a resounding yes. But, before we proceed, notice that the left-hand side of $yx^2 = 1$ is now a product of two factors: the implicit function $y = f(x)$ (*see bottom note*) and the explicit quantity x^2. Hence, when differentiating the expression $yx^2 = f(x) \cdot x^2$, we will need to use the standard differentiation process associated with the product rule. Continuing:

$$[yx^2]' = [1]' \Rightarrow$$
$$[y][x^2]' + [y]'[x^2] = 0 \Rightarrow$$
$$2xy + x^2 y' = 0 \Rightarrow$$
$$x^2 y' = -2xy \Rightarrow$$
$$y' = \frac{-2xy}{x^2} = \frac{-2y}{x}$$

In the above expression for y', we actually know what y is in terms of x since y is *solvable* in terms of x. Substituting $y = \dfrac{1}{x^2}$, we finally obtain $y' = \dfrac{-2}{x^3}$, which again matches the original y'.

Note: To reiterate, y is called an implicit function because we assume that a functional relationship of the form $y = f(x)$ is embedded in the expression $yx^2 = 1$. There is a powerful result found in advanced calculus, known as the Implicit Function Theorem, that stipulates the exact conditions for which this assumption is true. In this book, and in most standard calculus texts, we are going to proceed as if it were true.

If the previous three pages seem like unnecessary complexity for obvious and simple examples, consider the more complicated expression $2x + xy^3 + 1 = x^2 + y^2$. Here, we can't solve for y explicitly in terms of x. But, we can find y' using the techniques of implicit differentiation if we assume $y = f(x)$ and proceed accordingly using standard differentiation rules as now illustrated in **Ex 5.5.1**.

Ex 5.5.1: Find y' for $y = f(x)$ embedded in the defining expression $2x + xy^3 + 1 = x^2 + y^2$. When done, find the equation of both the tangent line and normal line at the point $(3,1)$.

Part 1: Pay careful attention to how both the product rule and generalized power rule are used under the governing assumption that y is indeed a function of x (i.e. $y = f(x)$). Hence, when differentiating y^3 and y^2, we have the following:

$$[y^3]' = [\{f(x)\}^3]' = 3\{f(x)\}^2 f'(x) = 3y^2 y'$$
$$[y^2]' = [\{f(x)\}^2]' = 2\{f(x)\}^1 f'(x) = 2yy'$$

Continuing:

$$2x + xy^3 + 1 = x^2 + y^2 \Rightarrow$$
$$[2x + xy^3 + 1]' = [x^2 + y^2]' \Rightarrow$$
$$[2x]' + [xy^3]' + [1]' = [x^2]' + [y^2]' \Rightarrow$$
$$2[x]' + x[y^3]' + [x]'y^3 + [1]' = [x^2]' + [y^2]' \Rightarrow$$
$$2 + 3xy^2 y' + 1 \cdot y^3 = 2x + 2yy' \mapsto$$
$$3xy^2 y' - 2yy' = 2x - 2 - y^3 \Rightarrow$$
$$(3xy^2 - 2y)y' = 2x - 2 - y^3 \Rightarrow$$
$$y' = \frac{2x - 2 - y^3}{3xy^2 - 2y}$$

Part 2: Evaluating y' at the point $(3,1)$, we obtain $y' = \frac{3}{7}$. The equation of the tangent line is given by $y - 1 = \frac{3}{7}(x - 3)$; the equation of the normal line, by $y - 1 = \frac{-7}{3}(x - 3)$.

Marvel at the power of this technique. Even though we don't know the explicit expression for y in terms of x, we can use *implicit differentiation* to find a viable y'. Subsequently, we can use this same y' to generate equations for tangent and/or normal lines given known points that satisfy the original expression. In many cases, this original expression can not be visualized via an actual graph unless done by math-enhancing software. The bottom line is that implicit differentiation greatly enhances the flexibility and applicability of the differentiation process in general. *One could say that implicit differentiation allows us to differentiate 'functions in hiding', hiding in either expressions or equations.*

Let's do two more tedious examples in order to firm up the implicit differentiation process before moving on to some practical applications involving *related rates*.

Ex 5.5.2: Find y' for $xy^6 - y^2 = x^3 + 10$

$$xy^6 - y^2 = x^3 + 10 \Rightarrow$$
$$[xy^6 - y^2]' = [x^3 + 10]' \Rightarrow$$
$$[xy^6]' - [y^2]' = [x^3]' + [10]' \Rightarrow$$
$$[x][y^6]' + [x]'[y^6] - [y^2]' = [x^3]' + [10]' \Rightarrow$$
$$6xy^5 y' + 1 \cdot y^6 - 2yy' = 3x^2 \Rightarrow$$
$$(6xy^5 - 2y)y' = 3x^2 - y^6 \Rightarrow$$
$$y' = \frac{3x^2 - y^6}{6xy^5 - 2y}$$

Ex 5.5.3: Find y' for $x^2 y^2 - \dfrac{x}{y} = 1 + y^2$

$$x^2y^2 - \frac{x}{y} = 1 + y^2 \Rightarrow$$

$$\left[x^2y^2 - \frac{x}{y} \right]' = [1 + y^2]' \Rightarrow$$

$$[x^2y^2]' - \left[\frac{x}{y} \right]' = [1]' + [y^2]' \Rightarrow$$

$$2x^2yy' + 2xy^2 - \left[\frac{y \cdot 1 - xy'}{y^2} \right] = 2yy' \Rightarrow$$

$$2x^2y^3y' + 2xy^4 - y + xy' = 2y^3y' \Rightarrow$$

$$2x^2y^3y' + xy' - 2y^3y' = y - 2xy^4 \Rightarrow$$

$$(2x^2y^3 + x - 2y^3)y' = y - 2xy^4 \Rightarrow$$

$$y' = \frac{y - 2xy^4}{2x^2y^3 + x - 2y^3}$$

Implicit differentiation can be immediately used in practical applications involving related rates. In related rates problems, several time-varying quantities are related in an algebraic expression or equation. Hence, each quantity within the equation or algebraic expression is a function of time, the whole of which can be differentiated using implicit differentiation. The derivatives produced are time-rates of change and can be algebraically related—origin of the term *related rates*—in order to solve a particular problem. Let's see how this process works in two practical applications

Ex 5.5.4: A spherical ball of ice is melting at the rate of $4\frac{in^3}{min}$. How fast is its radius changing when the radius is exactly 3 inches?

For a sphere, the formula relating volume to radius (Appendix B) is $V = \frac{4}{3}r^3$. Also, in the situation just described, both V and r are functions of time in minutes implying $V(t) = \frac{4}{3}[r(t)]^3$.

The expression $V(t) = \frac{4}{3}[r(t)]^3$ algebraically relates the two functions $V(t)$ and $r(t)$. Using implicit differentiation to differentiate the above expression, we obtain

$$V(t) = \tfrac{4}{3}[r(t)]^3 \Rightarrow$$
$$[V(t)]' = \left[\tfrac{4}{3}[r(t)]^3\right]' \Rightarrow$$
$$V'(t) = 4[r(t)]^2 r'(t)$$

The last expression relates $r(t)$ and the two *rates* $V'(t) \,\&\, r'(t)$. By the problem statement, we know that the ice ball is melting at a steady rate $V'(t) = 4\frac{in^3}{min}$. To find $r'(t)$ when $r(t) = 3in$, first solve the last equality for $r'(t)$ to obtain $r'(t) = \dfrac{V'(t)}{4[r(t)]^2}$. Now plug in the two specific values for r and V' in order to finish the job and (of course) answer the original question.

$$r'(t) = \frac{4\frac{in^3}{min}}{4[3in]^2} = \tfrac{1}{9}\tfrac{in}{min}.$$

Ex 5.5.5: A 20 foot ladder is leaning against a wall. A large dog is tied to the foot of the ladder and told to sit. Five minutes later, the dog is pulling the foot of the ladder away from the wall at a steady rate of $3\frac{ft}{sec}$, **Figure 5.26**. How fast is the top of the ladder sliding down the wall when the bottom is 14 feet from the wall?

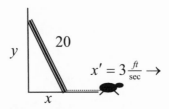

Figure 5.26: 'Large Dog' Pulling Ladder

By the Pythagorean Theorem, we have $x^2 + y^2 = 20^2$. Since x and y are both changing in time, $x = x(t)$ and $y = y(t)$. Substituting into the Pythagorean expression, we have that $[x(t)]^2 + [y(t)]^2 = 20^2$ which is our fundamental algebraic expression relating the rates. Differentiating (implicit style),

$$[x(t)]^2 + [y(t)]^2 = 20^2 \Rightarrow$$
$$\left[[x(t)]^2 + [y(t)]^2\right]' = [20^2]' \Rightarrow$$
$$2x(t)x'(t) + 2y(t)y'(t) = 0$$

We want $y'(t)$ when $x(t) = 14\,ft$ and $x'(t) = 3\frac{ft}{sec}$. Solving the last equation for $y'(t)$, we obtain $y'(t) = \dfrac{-x(t)x'(t)}{y(t)}$. Using the Pythagorean Theorem yet again, we have that $y(t) = 14.28\,ft$ at the instant $x(t) = 14\,ft$. Finally, substituting all the parts and pieces into the general expression for $y'(t)$, one

obtains $y'(t) = \dfrac{-14\,ft \cdot 3\frac{ft}{sec}}{14.28\,ft} = -2.94\frac{ft}{sec}$.

$$\int_a^b \overset{\cdot\cdot}{\cup} dx$$

Section Exercises

1. Use implicit differentiation to differentiate the following:

a) $x^2 + 3xy + y^2 = 4$ b) $\dfrac{x^3}{x+y} = 7y^2$ c) $\ln(x+y) = y$

2. Find the equations for both the tangent line and normal line to the graph of the expression $x^2 y + y = 2$ at the point $(1,1)$.

3. A person is walking towards a tower 150 feet high at a rate of $4\frac{ft}{s}$. How fast is the distance to the top of the tower changing at the instant the person is 60 feet away from the foot of the tower?

4. Two ships are steaming in the Gulf of Mexico. Ship A is precisely 40 nautical miles due north of ship B at 13:00 CST. Ship A is steaming due south at 15 knots and ship B is steaming due east at 5 knots. At what time will the closing distance between the two ships be a minimum? What is the minimum closing distance?

5.6) Higher Order Derivatives

Consider the function $f(x) = 4x^3 - 7x^2 + x + 11$ which can be differentiated to obtain $f'(x) = 12x^2 - 14x + 1$. Now, the newly-minted derivative $f'(x)$ is a function in its own right, a function that can be differentiated using the same processes as those applied to our original function f. Hence, we can define a *second derivative* $f''(x)$ for the original function $f(x)$ by the iterative process $f''(x) \equiv [f'(x)]'$. Going ahead and taking $f''(x)$, we obtain

$$f''(x) \equiv [f'(x)]' \Rightarrow$$
$$f''(x) = [12x^2 - 14x + 1]' \Rightarrow$$
$$f''(x) = [12x^2]' - [14x]' + [1]'.$$
$$f''(x) = 12[x^2]' - 14[x]' + 0$$
$$f''(x) = 24x - 14$$

Second derivatives can be denoted by either prime or differential notation. A listing of common notations follows:

1. $f''(x)$: read as f double prime of x
2. f'': read as f double prime
3. y'': read as y double prime

4. $\dfrac{d^2 y}{dx^2}$:read as "dee-two y by dee-two x"

or "dee-two y by dee x square"

One can iterate a second time in order to obtain a *third derivative* $f'''(x) = [f''(x)]' = [24x - 14]' = 24$. A third iteration gives $f^{(4)}(x) = f^{(IV)}(x) = [f'''(x)]' = [24]' = 0$, called a *fourth derivative*. Continuing in like fashion, any derivative higher than the fourth derivative is also zero.

Note: In many texts, Roman numerals are used to denote all fourth-order and higher derivatives. Parentheses are always used in conjunction with numerals in order to distinguish from exponents.

Ex 5.6.1: Find all orders of derivatives for $y = x^5 - 3x^4 + 7x^2 + 9$

$\overset{1}{\mapsto} : y = x^5 - 3x^4 + 7x^2 + 9 \Rightarrow$

$y' = [x^5 - 3x^4 + 7x^2 + 9]' \Rightarrow$

$y' = 5x^4 - 12x^3 + 14x$

$\overset{2}{\mapsto} : y'' = 20x^3 - 36x^2 + 14 \Rightarrow$

$y''' = 60x^2 - 72x \Rightarrow$

$y^{(4)} = 120x - 72 \Rightarrow$

$y^{(5)} = 120 \Rightarrow$

$y^{(6)} = 0 \Rightarrow y^{(7)} = 0 \Rightarrow \ldots$

As the above example demonstrates, the derivatives of polynomial functions decrease in algebraic complexity with an increase in order. The following two examples show the same is not true with rational and/or transcendental functions.

Ex 5.6.2: Find $f''(x)$ for $f(x) = \dfrac{x^2}{\sqrt{x^2 + 1}}$.

First, we must generate $f'(x)$.

$$\overset{1}{\mapsto} : f(x) = \frac{x^2}{\sqrt{x^2+1}} = \frac{x^2}{(x^2+1)^{\frac{1}{2}}} \Rightarrow$$

$$f'(x) = \frac{(x^2+1)^{\frac{1}{2}}[x^2]' - x^2[(x^2+1)^{\frac{1}{2}}]'}{[(x^2+1)^{\frac{1}{2}}]^2} \Rightarrow$$

$$f'(x) = \frac{(x^2+1)^{\frac{1}{2}} \cdot 2x - x^2 \cdot \frac{1}{2}(x^2+1)^{-\frac{1}{2}} \cdot 2x}{(x^2+1)} \Rightarrow$$

$$f'(x) = \frac{x^3+2x}{(x^2+1)^{\frac{3}{2}}}$$

Next, the newly-minted $f'(x)$ becomes the input to the derivative-taking process in order to generate $f''(x)$.

$$\overset{2}{\mapsto} : f''(x) = \left[\frac{x^3+2x}{(x^2+1)^{\frac{3}{2}}}\right]' \Rightarrow$$

$$f''(x) = \frac{(x^2+1)^{\frac{3}{2}}[x^3+2x]' - (x^3+2x)[(x^2+1)^{\frac{3}{2}}]'}{[(x^2+1)^{\frac{3}{2}}]^2} \Rightarrow$$

$$f''(x) = \frac{(x^2+1)^{\frac{3}{2}}(3x^2+2) - (x^3+2x) \cdot \frac{3}{2}(x^2+1)^{\frac{1}{2}} \cdot 2x}{(x^2+1)^3} \Rightarrow$$

$$f''(x) = \frac{(x^2+1)(3x^2+2) - (x^3+2x) \cdot 3x}{(x^2+1)^{\frac{5}{2}}} \Rightarrow$$

$$f''(x) = \frac{(x^2+1)(3x^2+2) - (x^3+2x) \cdot 3x}{(x^2+1)^{\frac{5}{2}}} \Rightarrow$$

$$f''(x) = \frac{2-x^2}{(x^2+1)^{\frac{5}{2}}}$$

Ex 5.6.3: Find $\dfrac{d^2y}{dx^2}$ for $y = x^3 e^{x^2}$.

$\overset{1}{\mapsto} : y = x^3 e^{x^2} \Rightarrow$

$\dfrac{dy}{dx} = (x^3)\dfrac{d}{dx}[e^{x^2}] + (e^{x^2})\dfrac{d}{dx}[x^3] \Rightarrow$

$\dfrac{dy}{dx} = (x^3) \cdot 2xe^{x^2} + (e^{x^2}) \cdot 3x^2 \Rightarrow$

$\dfrac{dy}{dx} = (x^3) \cdot 2xe^{x^2} + (e^{x^2}) \cdot 3x^2 \Rightarrow$

$\dfrac{dy}{dx} = x^2(2x^2 + 3)e^{x^2}$

****

$\overset{2}{\mapsto} : \dfrac{d^2y}{dx^2} = \dfrac{d}{dx}\left[\dfrac{dy}{dx}\right] = \dfrac{d}{dx}\left[x^2(2x^2+3)e^{x^2}\right] \Rightarrow$

$\dfrac{d^2y}{dx^2} = x^2(2x^2+3)\dfrac{d}{dx}[e^{x^2}] + e^{x^2}\dfrac{d}{dx}[x^2(2x^2+3)] \Rightarrow$

$\dfrac{d^2y}{dx^2} = x^2(2x^2+3) \cdot 2xe^{x^2} + e^{x^2} \cdot (8x^3 + 6x) \Rightarrow$

$\dfrac{d^2y}{dx^2} = 2x(2x^4 + 7x^2 + 3)e^{x^2}$

One can also use implicit differentiation to find a second-order derivative.

Ex 5.6.4: Find y'' for $x^2y + y^2x = 2y + 7x + 10$

$$\overset{1}{\mapsto}: x^2 y + y^2 x = 2y + 7x + 10 \Rightarrow$$
$$[x^2 y + y^2 x]' = [2y + 7x + 10]' \Rightarrow$$
$$[x^2 y]' + [y^2 x]' = 2[y]' + 7[x]' + [10]' \Rightarrow$$
$$2xy + x^2 y' + y^2 \cdot 1 + x \cdot 2yy' = 2y' + 7 \Rightarrow$$
$$2xy + x^2 y' + y^2 \cdot 1 + x \cdot 2yy' = 2y' + 7$$
$$(x^2 + 2xy - 2)y' = 7 - y^2 \Rightarrow$$
$$y' = \frac{7 - y^2}{x^2 + 2xy - 2}$$

We can continue with the above expression in order to find y'' ; but the previous expression lacks a denominator, which makes the algebra a little simpler.

Note: Mathematicians—if you can believe it—are basically a lazy lot always looking for ways to improve the process. Dr. C. E. Deming was a mathematician/statistician long before he became a champion of the American Quality Movement.

Continuing:

$$\overset{2}{\mapsto}: [(x^2 + 2xy - 2) \cdot y']' = [7 - y^2]' \Rightarrow$$
$$(x^2 + 2xy - 2) \cdot [y']' + [(x^2 + 2xy - 2)]' \cdot y' = [7]' - [y^2]' \Rightarrow$$
$$(x^2 + 2xy - 2) \cdot y'' + (2x + 2y + 2xy') \cdot y' = -2yy' \Rightarrow$$
$$(x^2 + 2xy - 2) \cdot y'' = -2yy' - 2xy' - 2yy' - 2x(y')^2 \Rightarrow$$
$$(x^2 + 2xy - 2) \cdot y'' = -4yy' - 2xy' - 2x(y')^2 \Rightarrow$$
$$y'' = \frac{-2y'(2y + x + xy')}{x^2 + 2xy - 2}$$

Finally, substituting $y' = \dfrac{7 - y^2}{x^2 + 2xy - 2}$, we obtain

$$\mapsto : y'' = \frac{-2(7-y^2)\left\{2y + x + x\left[\dfrac{7-y^2}{x^2+2xy-2}\right]\right\}}{(x^2+2xy-2)^2} \Rightarrow$$

$$y'' = \frac{-2(7-y^2)\{(2y+x)(x^2+2xy-2) + x(7-y^2)\}}{(x^2+2xy-2)^3} \Rightarrow$$

$$y'' = \frac{-2(7-y^2)\{x^3 + 4yx^2 + 3xy^2 - 4y + 5x\}}{(x^2+2xy-2)^3}$$

Now, for any function f having both first and second derivatives, the following fundamental syllogism applies:

The second derivative f'' is to f' as

The first derivative f' is to f

Hence, we can use f'' to analyze f' in exactly the same way that we used f' to analyze f. An immediate application is the finding of those points of the graph of f where the slope values (not to be confused with the functional values) have a local extrema. This simple observation leads to an equally simple definition.

Definition: A function f is said to have a *hypercritical point* at x_0 if either $f''(x_0) = 0$ or f'' does not exist at x_0.

Notice that both *critical points* and *hypercritical points* (*hypercritical points* are also known as *inflection points*) are functional attributes referenced to the original function f. This is made clear by the problem statement in our next example.

Ex 5.6.5: Find both the critical and hypercritical points *for the function* $f(x) = x^3 - 6x^2 + 9x$.

$$\overset{1}{\mapsto}: f(x) = x^3 - 6x^2 + 9x \Rightarrow$$
$$f'(x) = 3x^2 - 12x + 9 = 3(x-1)(x-3) \mapsto$$
$$f'(x) = 0 \Rightarrow 3(x-1)(x-3) = 0 \Rightarrow$$
$$x = 1, 3$$

The points $x = 1$ and $x = 3$ are *critical* points for f since $f' = 0$.

$$\overset{2}{\mapsto}: f''(x) = 6x - 12 = 6(x-2) \rightarrow$$
$$f''(x) = 0 \Rightarrow 6(x-2) = 0 \Rightarrow$$
$$x = 2$$

The point $x = 2$ is a *hypercritical* (or *inflection*) point for f since $f'' = 0$.

Let's take **Ex 5.6.5** one step further by giving it a physical context. Suppose a roller coaster is moving on the graph of $f(x) = x^3 - 6x^2 + 9x$ as shown in **Figure 5.27**.

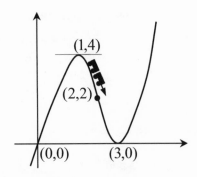

Figure 5.27: A Roller Coaster Ride

After rounding the peak at $(1,4)$, the coaster dives towards the low point at $(3,0)$ (*note: the reader should verify that* $(1,4)$ *is a local maximum and* $(3,0)$ *is a local minimum*). The dive angle initially steepens as the coaster descends, causing weak stomachs to fly.

However, this steepening must cease about midway into the dive; or else, the coaster will forcibly bury itself at the bottom. Consequently, there is a point, the *hypercritical point* at $(2,2)$, on the dive path where the coaster starts pulling out of the dive in order to end up with a horizontal slope at $(3,0)$. Thus, the hypercritical point $(2,2)$ marks two important events on our coaster ride: 1) the location of maximum steepness and 2) where the much-needed pull-out begins. Finally, if we were climbing from a low point to a high point, a similar analysis would apply. There would have to be a point in the climb path where we would need to start a leveling process in order to become horizontal at the high point.

To finish Section 5.6, let's briefly explore the relationship between derivatives of various orders and simple motion. Suppose the horizontal displacement, in reference to a starting point, of an object after t hours is given by $P(t) = 5t + 10t^2$ where $P(t)$ is in feet. We have already seen that the first derivative $P'(t) = 5 + 20t$ has units $\frac{feet}{hour}$ and can be interpreted in this motion context as *instantaneous velocity*. Likewise, the second derivative $P''(t) = 20$ has units $\frac{\frac{feet}{hour}}{hour} = \frac{feet}{hour^2}$ and can be interpreted as *instantaneous acceleration*. Recall from physics that acceleration is defined as the time-rate of change of velocity. The third derivative $P'''(t) = 0$ and has units $\frac{feet}{hour^3}$. The third derivative defines a quantity called *instantaneous jerk*. Jerk is well named; for when jerk is non-zero, one will describe the associated motion as jerky or non-smooth.

Ex 5.6.6: A hammer is dropped from the roof of the Sears Tower in Chicago, Illinois. Find the impact velocity and acceleration at the time of impact if $P(t) = 1450 - 16t^2$ gives hammer height above street level t seconds after hammer release.

To find the total elapsed time from hammer release until hammer impact, set $P(t) = 0$.

$$P(t) = 1450 - 16t^2 = 0 \Rightarrow$$
$$1450 - 16t^2 = 0 \Rightarrow$$
$$t = 9.51\,\text{sec}$$

To find hammer velocity and acceleration at impact, evaluate

$$P'(t) = -32t \text{ and } P'(9.51) = -304.63\,\tfrac{ft}{\text{sec}}$$
$$P''(t) = -32 \text{ and } P''(9.51) = -32\,\tfrac{ft}{\text{sec}^2}$$

$$\int_a^b \ddot{\cup}\,dx$$

Section Exercises

1. Find y' and y'' for the following functions:

a) $y = 7x^4 - 5x^2 + 17$ b) $y = \dfrac{x^2}{\sqrt{x^2 + 9}}$ c) $y = xe^x - x\ln x$

2. Find y'' for $y = xy^2 + x^2 + 2$.

3. The point $(2,3)$ is on the ellipse defined by $\dfrac{x^2}{8} + \dfrac{y^2}{18} = 1$. Find the equations for both the tangent line and normal line.

4. How far will a truck travel after the brakes are applied if the braking equation of forward motion is given by $D(t) = 100t - 10t^2$, where $D(t)$ is in feet and t is in seconds?

5.7) Further Applications of the Derivative

5.7.1) The Second Derivative Test

Definition: A function f is called *twice differentiable* on an interval (a,b) if one can generate both f' and f'' for every point within the interval.

Now suppose f is twice differentiable in (a,b) and there is a *critical point* x_0 within (a,b) where $f'(x_0) = 0$. Then, by the fundamental differential relationship, we have the following result:

$$f'(x_0 + dx) - f'(x_0) = f''(x_0)dx \mapsto$$
$$f'(x_0) = 0 \Rightarrow f'(x_0 + dx) = f''(x_0)dx$$

The expression $f'(x_0 + dx) = f''(x_0)dx$ forms the basis for the second-derivative test. This test distinguishes a local maximum from a local minimum.

Second Derivative Test for Local Extrema

Let f be twice differentiable in (a,b) and let x_0 be a *critical point* within (a,b). *Note: the fact that f is differentiable in (a,b) means that x_0 will be restricted to the type of critical point where $f'(x_0) = 0$*. Then:

Case 1: $f''(x_0) > 0$ implies f has a local minimum at x_0.
Case 2: $f''(x_0) < 0$ implies f has a local maximum at x_0.
Case 3: $f''(x_0) = 0$ means the test can't be used.

$f''(x_0) = 0$ also means that we must resort to the first derivative test if we need the information bad enough.

To prove the second derivative test, the expression $f'(x_0 + dx) = f''(x_0)dx$ is repeatedly used for various sign combinations of dx and $f''(x_0)$ as we progress through the three cases. Overall conclusions are captured visually via three generic sign-change charts.

$1: for \quad f''(x_0) > 0 : dx > 0 \Rightarrow f'(x_0 + dx) = f''(x_0)dx > 0$

$and \quad dx < 0 \Rightarrow f'(x_0 + dx) = f''(x_0)dx < 0$

$\mapsto : ----[\overset{\cup}{\underset{0}{x_0}}]++++ \therefore$

$2: for \quad f''(x_0) < 0 : dx > 0 \Rightarrow f'(x_0 + dx) = f''(x_0)dx < 0$

$and \quad dx < 0 \Rightarrow f'(x_0 + dx) = f''(x_0)dx > 0$

$\mapsto : ++++[\overset{\cap}{\underset{0}{x_0}}]---- \therefore$

$3: for \quad f''(x_0) = 0 : \forall dx \Rightarrow f'(x_0 + dx) = f''(x_0)dx = 0$

$\mapsto ????[\overset{?}{\underset{0}{x_0}}]???? \therefore$

Thus, the test fails miserably in 3 (Case 3)

Note: All is not lost if $f'(x_0) = f''(x_0) = 0$. Double zeros are a strong indicator of a saddle point. However, we will still need to perform the first derivative test in order to definitively check out our hunch.

Ex 5.7.1: Find the local extrema for $f(x) = 2x^3 - 9x^2 - 24x + 1$.

This example is the same as **Ex 5.4.7**. But here, we modify Step 2 in order to incorporate the second derivative test.

$\overset{1}{\mapsto} : f'(x) = 6x^2 - 18x - 24 = 6(x-4)(x+1) \mapsto$

$f'(x) = 0 \Rightarrow 6(x-4)(x+1) = 0 \Rightarrow x = 4, -1$

$$\overset{2}{\mapsto}: f''(x) = 12x - 18 \mapsto$$
$$f''(-1) = 12(-1) - 18 = -30 < 0 \Rightarrow \cap$$
$$f''(4) = 48 - 18 = 30 > 0 \Rightarrow \cup$$

****

$$\overset{3}{\mapsto}: f(x) = 2x^3 - 9x^2 - 24x + 1 \Rightarrow$$
$$f(-1) = 14 \ \& \ f(4) = -111$$

In order to complete the job, we must evaluate f at the critical points -1 and 4 (see Subsection 5.4.3).

Ex 5.7.2: Show that $e^\pi > \pi^e$.

Note: I first encountered this devilish problem while taking a college algebra class (fall of 1965) and have never been able to solve it using just algebraic techniques. In 1965, the slide rule was still king. So, no fair plugging numbers into your hand-held graphing marvel. Calculus finally came to my rescue some 20 years later. The limerick at the end of the chapter commemorates this victory via a mathematical challenge.

Define $f(x) = e^x - x^e$ on the interval $[0,10]$, an interval for which f is twice differentiable. Notice that $\pi \in (0,10)$ and $f(\pi) = e^\pi - \pi^e$ is the quantity to be examined.

$$\overset{1}{\mapsto}: \quad \text{Find the critical points for } f.$$

$$f(x) = e^x - x^e \Rightarrow f'(x) = e^x - ex^{e-1} \rightarrow$$
$$f'(x) = 0 \Rightarrow e^x - ex^{e-1} = 0 \Rightarrow$$
$$e^x = ex^{e-1} \Rightarrow \ln(e^x) = \ln(ex^{e-1}) \Rightarrow$$
$$x = \ln(e) + (e-1)\ln(x) \Rightarrow$$
$$x = 1, e$$

$\overset{2}{\mapsto}:$ Apply the second derivative test.

$$f''(x) = e^x - e(e-1)x^{e-2}$$
$$f''(1) = e - e(e-1) = e(2-e) < 0 \Rightarrow \cap$$
$$f''(e) = e^e - e(e-1)e^{e-2} = e^{e-1} > 0 \Rightarrow \cup$$

$\overset{3}{\mapsto}:$ Interpret the results. Now $f(e) = 0$ is the only local minimum on $[0,10]$.   Since $f(0) = 1 > 0$, $f(1) = e - 1 > 0$ and $f(10) = e^{10} - 10^e > 0$, we also have that $f(e) = 0$ is an absolute minimum on $[0,10]$. Hence, we conclude $f(x) > 0$ for all other $x \in (0,10)$ and, in particular, for $x = \pi$. Thus:

$$f(\pi) > 0 \Rightarrow$$
$$f(\pi) = e^\pi - \pi^e > 0 \Rightarrow$$
$$e^\pi > \pi^e \therefore$$

Ex 5.7.3: Find the local extrema for $f(x) = x^3 - 3x^2 + 3x + 4$.

$\overset{1}{\mapsto}: f'(x) = 3x^2 - 6x + 3 = 3(x-1)^2 \mapsto$
$$f'(x) = 0 \Rightarrow 3(x-1)^2 = 0 \Rightarrow x = 1$$
$$****$$

$\overset{2}{\mapsto}: f''(x) = 6x - 6 \mapsto$
$$f''(1) = 6(1) - 6 = 0$$

The second derivative test fails since $f''(1) = 0$. Therefore, we must use the first derivative test in order to complete the analysis.

$\overset{3}{\mapsto}: f'(x) = 3(x-1)^2 \mapsto$
$$\overset{S}{+ + + + [1] + + + +}_{0}$$

The critical point at $x = 1$ is revealed by the first derivative test to be a saddle point with $f(1) = 5$. This supports the previous comment on double zeros.

Ex 5.7.4: Find the local extrema for $f(x) = x + \dfrac{1}{x}$.

$$\overset{1}{\mapsto} : f'(x) = 1 - \frac{1}{x^2} = \frac{x^2 - 1}{x^2} \mapsto$$

$$f'(x) = 0 \Rightarrow \frac{x^2 - 1}{x^2} = 0 \Rightarrow x = 1, -1$$

****

$$\overset{2}{\mapsto} : f''(x) = \frac{2}{x^3} \mapsto$$

$$f''(-1) = -2 < 0 \Rightarrow \cap$$

$$f''(1) = 2 > 0 \Rightarrow \cup$$

****

$$\overset{3}{\mapsto} : f(x) = x + \frac{1}{x} \Rightarrow$$

$$f(-1) = -2 \ \& \ f(1) = 2$$

5.7.2) Geometric Optimization with Side Constraints

Sometimes an optimization problem will require a primary quantity to be optimized when a related quantity is being constrained. **Ex 5.4.16** can be thought of in these terms: a volume of an open-topped box is to be maximized under the condition that a square piece of paper is to be constrained to a side length of 10 inches. Viewing an optimization problem in terms of constraints can often be a viable approach in solving word problems where several different expressions are related to the quantity to be optimized. Two examples will illustrate the process.

Ex 5.7.5: A closed rectangular box is to be constructed that has a volume of $4\,ft^3$. The length of the base of the box will be twice as long as its width.

The material for the top and bottom of the box costs $0.30 per square foot. The material for the sides of the box costs $0.20 per square foot. Find the dimensions of the least expensive box that can be constructed. What is the cost of such a box?

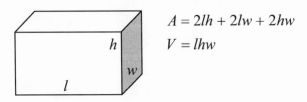

$$A = 2lh + 2lw + 2hw$$
$$V = lhw$$

Figure 5.28: Enclosed Rectangular Box

Figure 5.28 shows an enclosed rectangular box with three labeled dimensions. The cost of producing such a box irregardless of size will be

$$C(l,h,w) = (\$0.20)2lh + (\$0.30)2lw + (\$0.20)2hw \Rightarrow$$
$$C(l,h,w) = \$0.40lh + \$0.60lw + \$0.40hw$$

One immediately notices that $C = C(l,h,w)$ is a function of three independent variables, as opposed to a function of a single independent variable. Thus, the first order of business is to reduce the number of independent variables from three down to one, so we can apply the *single-variable optimization techniques* as found in this chapter.

From the problem statement, we have that *the length of the base of the box will be twice as long as its width*. This condition is a *side constraint,* which is easily captured by the expression $l = 2w$. Hence, the independent variable l is not so independent after all. Consequently, the substituting of $l = 2w$ into the expression for cost leads to a reduction in the number of independent variables.

$$C(l,h,w) = \$0.80wh + \$1.20w^2 + \$0.40hw = C(h,w) \Rightarrow$$
$$C(h,w) = \$1.20wh + \$1.20w^2$$

Two independent variables is still one too many. To reduce further, we employ the additional *side constraint* $V = lwh = 4\,ft^3$, which leads to the following:

$$lwh = 4 \Rightarrow 2w^2h = 4 \Rightarrow h = \frac{2}{w^2} \mapsto$$

$$C(h,w) = \$1.20wh + \$1.20w^2 \Rightarrow$$

$$C(h,w) = \$1.20w\left[\frac{2}{w^2}\right] + \$1.20w^2 = C(w) \Rightarrow$$

$$C(w) = \frac{\$2.40}{w} + \$1.20w^2$$

With the help of two side constraints, we have reduced $C(l,h,w)$ to the function $C(w)$, a function having a single independent variable. Notice that in the context of the physical problem, the function $C(w)$ has natural domain $(0,\infty)$, much of which is unusable since costs become extremely large as $w \rightarrow 0$ or $w \rightarrow \infty$. So, hopefully, there is a 'sweet w' somewhere in the interval $(0,\infty)$, which is of reasonable size and minimizes the cost of construction. Formulation complete, we are now ready to continue with the easier pure-math portion of the problem.

$$\overset{1}{\mapsto} : C'(w) = -\frac{2.40}{w^2} + 2.40w \mapsto$$

$$C'(w) = 0 \Rightarrow w = 1$$

****

$$\overset{2}{\mapsto} : C''(w) = \frac{4.80}{w^3} + 2.40 \mapsto$$

$$C''(1) = 7.20 > 0 \Rightarrow \cup$$

****

$\overset{3}{\mapsto}$: Answer the two questions.

$$w = 1\,ft \Rightarrow l = 2w = 2\,ft \Rightarrow h = \frac{2}{w^2} = 1\,ft$$

$$C(w) = \frac{\$2.40}{w} + \$1.20w^2 \Rightarrow C(1) = \$3.60$$

Ex 5.7.6: Of all rectangles with given perimeter, which has maximum area? What is the maximum area so obtained?

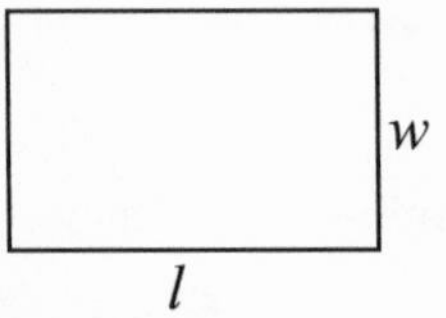

$$A = lw$$
$$P = 2l + 2w$$

Figure 5.29: Rectangle with Given Perimeter

Figure 5.29 shows the layout of this rather elementary, but fundamental, example. We are to maximize $A(l, w) = lw$ subject to the *side constraint* $P = 2l + 2w$. The first task is to reduce the number of independent variables as before:

$$P = 2l + 2w \Rightarrow l = \frac{P - 2w}{2} \Rightarrow$$

$$A(l, w) = \left[\frac{P - 2w}{2} \right] w = A(w) \Rightarrow$$

$$A(w) = \frac{Pw - 2w^2}{2} : w \in [0, \tfrac{P}{2}]$$

All areas corresponding to points in $(0, \tfrac{P}{2})$ are positive. Also, notice that $A(0) = A(\tfrac{P}{2}) = 0$, which corresponds to the using of all available perimeter on width.

$$\overset{1}{\mapsto}: A'(w) = \tfrac{1}{2}[P - 4w] \mapsto$$
$$A'(w) = 0 \Rightarrow w = \tfrac{P}{4}$$
****

$$\overset{2}{\mapsto}: A''(w) \equiv -2 < 0 \Rightarrow \cap$$
****

$$\overset{3}{\mapsto}: w = \tfrac{P}{4} \Rightarrow l = \tfrac{P}{4} \Rightarrow A(\tfrac{P}{4}) = \frac{P^2}{16}$$

Conclusion: the rectangle enclosing the greatest area for a given perimeter P is a square of side $\tfrac{P}{4}$. The associated area is $\dfrac{P^2}{16}$.

$$\int_a^b \overset{\bullet\bullet}{\cup} dx \quad \int_a^b \overset{\bullet\bullet}{\cup} dx$$

E to the Pi beats Pi to the e.
I claim it so! Do you agree?
The calculus rules
But algebra fools—
No fun if done with keystroke and key!

October 2001

Chapter Exercises

1. The U.S. Postal Service will not accept rectangular packages having square ends if the girth (perimeter around the package) plus the length exceeds 108 inches. What are the dimensions of an acceptable package having the largest possible volume? What is the associated volume?

2. Small spherical balloons are being filled with helium at a steady volumetric flow rate of $50\frac{in^3}{s}$ via a commercial gas injector. One particular brand of balloon will pop if the total surface area exceeds $144 in^2$. How many seconds do you have to remove this particular brand of balloon once the fill volume reaches $100 in^3$?

3. Find the local extrema for each of the following functions.

a) $f(x) = (x-1)x^{\frac{2}{3}}$ b) $f(x) = \dfrac{x}{x^2+1}$

c) $f(x) = x^2 e^x$

4. Find the equation of the tangent line at the point $(2,-1)$ for the function implicitly defined by $2y + 5 = x^2 + y^3$. Find the equation of the normal line at the same point. Finally, find y''.

5. A box with a square base and an open top is to be made with $27\,ft^2$ of cardboard. What is the maximum volume that can be contained within the box?

6. Find the absolute extrema for $f(x) = x^3 + 3x^2 - 24x$ on $[0,3]$.

Newton's Whit

Within the world of very small
Exists the tiniest whit of all,
One whose digits add no gain
To a nit or single grain;
And if a whit measures snow,
Add one flake to winter's toll.

Even with size so extreme,
Divisible still is scale by scheme;
For whit over whit tallies well
Numbering a world with much to tell:
From optimum length to girth of stars,
From total lift to time to Mars.

And thus we tout Sir Isaac's whit
Praising both beauty and benefit,
Yet, ol' Leibniz can claim...
A good half of it!

September 2004

6) Antiprocesses

"To every action there is always opposed an equal reaction:
Or, the mutual actions of bodies upon each other
Are always equal, and directed to contrary parts"
Sir Isaac Newton

6.1) Antiprocesses Prior to Calculus

When Isaac Newton made the above statement, he was speaking about a principle in the realm of physics linking mass, force, and motion. In the realm of mathematics, we have a similar principle; and, mimicking Newton, we can state this principle in terms of processes.

> For every mathematical process, there is always an associated *antiprocess*: which, by definition, is a new mathematical process that reverses the action of the original process.

Our first encounter with this principle came in 2nd or 3rd grade *when we started to learn subtraction (sometimes called take away) which undoes the process of addition.* Our second encounter was when we started to learn division which undoes the process of multiplication. This process-antiprocess pedagogy continued to follow us into algebra, where the general educational pattern was always the same: first introduce the process and then, the associated antiprocess. In this book, we introduced the concept of function—as a particular type of process—in Section 4.1. Inverse functions (*one could call these antifunctions*) followed in Section 4.2.

If you reflect back on your mathematics education to date, you probably realize that the antiprocess, in most instances, is a little more difficult to master than the original process. A somewhat strained metaphor, but I liken the performing of an antiprocess to reassembling a shattered Humpty Dumpty (**Figure 6.1**) after performing the *forward process* of shoving him off the wall.

This is because forward processes tend to be straightforward and, as a rule, break the problem down into successive little steps where each step is easily accomplished given enough understanding and practice. Antiprocesses start with the outputs produced by associated forward processes and attempt to recreate the original inputs. This recreation can be extremely difficult since the output produced by a forward process is usually algebraically simplified to the point that the steps that led to the output are obscured. Hence, antiprocesses require more pattern recognition and intuition when being applied.

> *Humpty Dumpty sat on a wall,*
> *Humpty Dumpty had a great fall;*
> *All the King's horses and all the King's men*
> *Could not put Humpty Dumpty together again.*

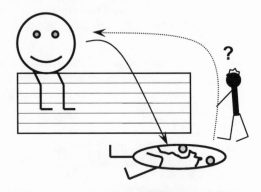

Figure 6.1: Poor Old Humpty Dumpty

Below is a small table listing some elementary processes from arithmetic and algebra with their associated antiprocesses.

Subject	Process	Antiprocess
Arithmetic	Addition	Subtraction
Arithmetic	Multiplication	Division
Algebra	Building Fractions	Reducing Fractions
Algebra	Polynomial Multiplication	Polynomial Factoring
Algebra	Adding Fractions	Partial Fractions
Algebra	Function	Inverse Function

Table 6.1: Selected Processes and Antiprocesses

The list in **Table 6.1** is not meant to be all inclusive. It is only to show that the process-antiprocess idea has been a part of our mathematics education for a long time—even though we may not have verbalized it as such in textbooks.

We end this section with some elementary factoring and check problems. Why? Factoring is a crucial algebraic skill that illustrates the process-antiprocess idea in a concise fashion. The forward process, polynomial multiplication, has set algebraic procedures requiring very little pattern recognition and/or intuition. However, the reverse process—polynomial factoring—is much more subtle. Polynomial factoring requires subtle pattern-recognition skills, obtainable only through practice, as we attempt to reconstruct the individual factors from which the polynomial product has been made. Re-examining polynomial multiplying and factoring from a process-antiprocess viewpoint is a great warm up for our next major section, which introduces the subject of *Antidifferentiation*.

$$\int_a^b \ddot{\cup}\, dx$$

Section Exercise

Factor (antiprocess) the following polynomials and check by multiplying (forward process). For each problem, reflect on the relationship between the two processes and the relative degree of difficulty. *Note: Appendix C has a list of factor formulas.*

a) $x^2 + 14x + 49$ b) $x^3 + x^2 + 3x + 3$ c) $3x^4 - 48$

d) $4x^2 + 16$ e) $6y^2 - 8y^3 + 4y^4$ f) $36x^2 - 25$

g) $6x^2 - 5x + 1$ h) $x^2 - x - 12$

i) $8x^3 + 22x^2 - 6x$ j) $3x^2 - 10x - 8$

k) $3m^2 - 9mn - 30n^2$ l) $25x^2 - 20x + 4$

6.2) Process and Products: Antidifferentiation

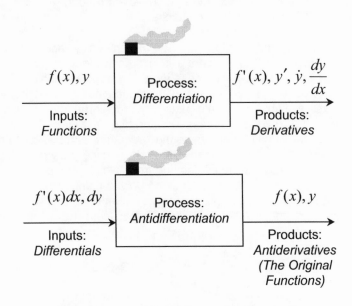

**Figure 6.2: The Process of Differentiation
Shown with the Process of Antidifferentiation**

Figure 6.2 shows the process of antidifferentiation along with the forward process of differentiation (see **Figure 5.3**). Antidifferentiation is the process by which we reconstruct the original function (called an antiderivative) from the associated differential (i.e. put Old Humpty Dumpty back together again). Together, differentiation and antidifferentiation comprise the two main processes of calculus.

Each of the two processes has its own historical conventions and corresponding symbols. In differentiation, the prime notation $[f(x)]'$ means to find the derivative for the function $f(x)$ and can also be used to denote the finite product as $f'(x)$. Hence the fundamental equality $[f(x)]' = f'(x)$ makes perfectly good sense because it states that the result from the differentiation process (left hand side) is the derivative as a product (right hand side).

Likewise, antidifferentiation comes with its own processing symbols and conventions. In antidifferentiation, we start the function reconstruction process with differentials as opposed to derivatives. Recall that if $y = f(x)$, associated differentials, differential change ratios, and derivatives are tightly interlinked and related by the foundational two-sided implication

$$\frac{dy}{dx} = f'(x) \Leftrightarrow dy = f'(x)dx.$$

This makes any one of the three aforementioned quantities an appropriate starting point for reconstruction of the original function $y = f(x)$. The traditional starting point is the quantity $f'(x)dx$. The rationale for this choice will become readily apparent in the next chapter as we solve the **Second Fundamental Problem of Calculus**.

The symbol used to annotate the antidifferentiation or function reconstruction process is the traditional *long S* symbol $\int$ which is called an integration (or integral) sign—not an antidifferentiation sign.

Note: The injection of this additional 'integration' terminology calls for an explanation. Historically, antidifferentiation has been known as indefinite integration. The term integration is a very appropriate term describing function reconstruction in that it conveys the idea of function reassembly from various parts and pieces. In today's engineering world, the term integration is used extensively and means the formation of a viable, interacting engineering system from component parts. Hence, integration is a good modern word. But antidifferentiation, a late twentieth-century term, is more suggestive of the actual function reconstruction process being performed, especially in terms of inputs to the process. Hence, it is also a good word. What do we do in mathematics when we have two good words, one historical and one descriptive? Answer: Keep them both and use them both to describe the same process.

In practice the symbol $\int f'(x)dx$ tells us to find a function whose differential is $f'(x)dx$. Simple intuition would immediately recognize one such function as $f(x)$.

Thus, one could be tempted to write $\int f'(x)dx = f(x)$ as a correct description of the antidifferentiation process.

But is this totally correct? Not quite. Examine the expression $\int x^2 dx = \dfrac{x^3}{3}$. Here, the differential is $x^2 dx = dy$ and the reconstructed function is $\dfrac{x^3}{3} = f(x) = y$. By definition, the function $f(x) = \dfrac{x^3}{3}$ certainly is an antiderivative for $x^2 dx$ since $f'(x) = x^2$. However, it is not the only one. The function $f(x) = \dfrac{x^3}{3} + 7$ is also an antiderivative for $x^2 dx$. In fact, any function of the general form $f(x) = \dfrac{x^3}{3} + C$ where C is an arbitrary constant qualifies as an antiderivative for $x^2 dx$. Hence, a far more correct answer when evaluating $\int x^2 dx$ is $\int x^2 dx = \dfrac{x^3}{3} + C$, where C is understood to be an arbitrary constant.

We shall now state and prove a fundamental theorem which will lead immediately to the basic antidifferentiation formula.

Let $f(x)$ and $g(x)$ be such that $f'(x) = g'(x)$
for all x values in an open interval (a,b).
Then we have $f(x) = g(x) + C$ for all x values in (a,b).

The theorem states that if two functions have identically matching derivatives on an interval, then the functions themselves must match to within a constant on the same interval.

The proof is very simple.

Define $F(x) = f(x) - g(x)$ on (a,b) and differentiate F. We have that $F'(x) = f'(x) - g'(x) = 0$ for all x values in (a,b). Therefore the function F itself (since the slope is identically zero) must be a horizontal line parallel to the x axis. This means F has the general functional form $F(x) = C$ which implies $f(x) = g(x) + C$ ∴.

Figure 6.3 illustrates the last result by depicting several functions from a notational functional family where each function has the general form $f(x) + C$. All functions in this family have identical slope behavior as given by $f'(x)$ but different y intercepts, which correspond to various selected values for the arbitrary constant C. An important point to remember is that we can not escape the functional family once our derivative is known. For if g is any function whatsoever with $g'(x)$ equal to $f'(x)$; then $g(x)$, by necessity, also has the form $f(x) + C$.

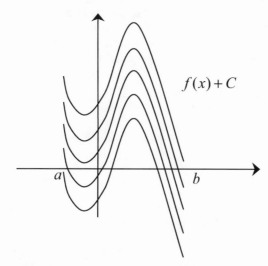

Figure 6.3: The Functional Family Defined by $f'(x)$

Basic Antidifferentiation Formula

Let $F(x)$ be such that $F'(x) = f(x)$.

If C is an arbitrary constant, then $F(x) + C$ represents the complete family of antiderivatives for $f(x)$ and

$$\int f(x)dx = F(x) + C.$$

In reference to the Basic Antidifferentiation Formula, let's discuss the symbol hierarchy used in calculus. The two basic calculus processes, differentiation and antidifferentiation, are performed on functions. In the case of antidifferentiation, these functions serve as derivatives. But, nonetheless, they are still functions by definition and are typically given the generic input name $f(x), g(x)$, etc. If $f(x)$ is serving as an input to the differentiation process, we will represent the output (derivative) by the prime notation $f'(x)$. However, if the same $f(x)$ is serving as part of an input to the antidifferentiation process, we will represent the output (antiderivative) by the capital-letter notation $F(x)$. The prime/capital-letter notation is the standard symbol hierarchy for differentiation and antidifferentiation and was first used when stating the Basic Antidifferentiation Formula. This standard notation is shown diagrammatically in **Figure 6.4**.

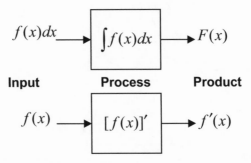

Figure 6.4: Annotating the Two Processes of Calculus

Now, let's use the Basic Antidifferentiation Formula to work four relatively easy antidifferentiation examples.

Ex 6.2.1: Find $G(x)$ for $g(x) = x^2 + 1$ and check your answer.

The associated differential for the function $g(x)$ is $(x^2 + 1)dx$ and the symbol $\int (x^2 + 1)dx$ means to conduct the antidifferentiation process. At this point in the book, we do not have a set of handy antidifferentiation rules to help us obtain the answer. Thus, our method for this example will be educated guessing. *Note: educated guessing is also used in many instances to perform trinomial factoring.* Our educated guess is

$$\int (x^2 + 1)dx = \frac{x^3}{3} + x + C \Rightarrow G(x) = \frac{x^3}{3} + x + C .$$

So, how does one verify the above guess? Simply differentiate $G'(x)$. If we have $G'(x) = g(x)$, then you have the right answer. For this example $G'(x) = \frac{1}{3}[3x^2] + 1 + 0 = x^2 + 1 = g(x)$, which is a match.

Note: checking an antidifferentiation problem is akin to checking a factoring problem. In factoring, you multiply the factors obtained in hopes of replicating the original expression. In antidifferentiation, you differentiate the antiderivative obtained in hopes of obtaining the original function. Unfortunately, checking factoring is far easier than doing factoring. The same is true for antidifferentiation!

Ex 6.2.2: Find $F(x)$ for $f(x) = x^3 + 2x^2 + x + 5$ and check your answer.

Again, we use educated guessing to obtain

$$F(x) = \int (x^3 + 2x^2 + x + 5)dx = \frac{x^4}{4} + \frac{2x^3}{3} + \frac{x^2}{2} + 5x + C$$

The answer check is left to the reader.

Before continuing with our last two examples, start noticing the general patterns associated with antidifferentiation in the first two examples, particularly when we have antidifferentiated powers, sums of terms, and constant multipliers. These general patterns will mimic the ones found in differentiation. For example, in differentiation, we reduce a power by one. In antidifferentiation, we increase a power by one. In Section 6.3, we present a complete set of elementary antidifferentiation rules—rules also commonly known as integral formulas or integration formulas—that will put some systematic process into the educated guessing. But for now, we will still guess.

Ex 6.2.3: Find $F(x)$ for $f(x) = \dfrac{2x}{x^2+1}$ and check your answer.

Solving: $F(x) = \displaystyle\int \frac{2x}{x^2+1}dx = \ln(x^2+1) + C$

Checking:
$$F(x) = \ln(x^2+1) + C \Rightarrow$$
$$F'(x) = \frac{[x^2+1]'}{x^2+1} = \frac{2x}{x^2+1} = f(x)$$

Ex 6.2.4: Find $F(x)$ for $f(x) = \sqrt{x} + 2$ such that $F(1) = 3$. For this same function, find $F(0)$.

$\overset{1}{\mapsto}$: (My guess) $F(x) = \displaystyle\int(\sqrt{x}+2)dx = \frac{2\sqrt{x^3}}{3} + 2x + C$

The equality $F(1) = 3$ is called a boundary or constraining condition and is used to precisely lock in the value for C as shown in Step 2.

$$\overset{2}{\mapsto} : F(1) = 3 \Rightarrow$$

$$\frac{2\sqrt{1^3}}{3} + 2(1) + C = 3 \Rightarrow \frac{8}{3} + C = 3 \Rightarrow C = \frac{1}{3} \Rightarrow$$

$$F(x) = \frac{2\sqrt{x^3}}{3} + 2x + \frac{1}{3}$$

With C now precisely determined, we can answer the last question as shown in Step 3.

$$\overset{3}{\mapsto} : F(0) = \frac{2\sqrt{0^3}}{3} + 2(0) + \frac{1}{3} = \frac{1}{3}$$

$$\int_a^b \overset{\bullet\bullet}{\cup} dx$$

No Exercises in This Section

6.3) Process Improvement: Integral Formulas

Integral formulas are the reverse of derivative formulas. Integral formulas are also called integration formulas, antidifferentiation formulas, or antiderivative formulas (all four terminolgies are used via mix-or-match style in modern textbooks). The terms integral and antiderivative place the emphasis on the product whereas the terms integration or antidifferentiation place the emphasis on the process. This book also uses a mixed terminology so that you become comfortable with all the words in the antidiiferentiation vocabulary.

Since integration (or antidifferentiation) is the antiprocess for differentiation, it makes sense that many of the derivative formulas in Section 5.3 can be reversed in order to obtain the appropriate integral formula.

This will be the procedure used in this volume. Hence, we will not prove most of the integral formulas in Section 6.3 since such a proof would be nothing more than a reversal of the proof for the associated derivative formula.

One final point is repeated from the Humpty Dumpty example, "...all the King's men could not put Humpty Dumpty together again." The same can happen when doing antidifferentiation. In this book, we will present a few basic antidifferentiation formulas and illustrate their use. But, be warned, it doesn't take a whole lot of imagination to create an algebraic expression that is very difficult (if not impossible) to antidifferentiate. An immediate example, where it is not easy to find an antiderivative $F(x)$ with $F'(x)$ equal to a given $f(x)$, is the

simple-to-look-at expression $f(x) = \sqrt{1 - x^4}$. In this book, we steer clear of advanced techniques and tricks; also, simple-to-look-at examples whose antiderivaties embody functions not covered in this volume (e.g. trigonometric and hyperbolic functions). To include these methods would require a volume about double the size of the one that you are currently holding and more formulas than you can imagine.

Note: While writing this section, I have before me a small out-of-print book entitled A Short Table of Integrals by B.O. Peirce, dated 1929. My father used it while studying radar at Harvard just after the start of WWII. There are 938 integral formulas in this book, and this is the short version. I have personally seen as many as 3000 integral formulas in a book. With all these 'instantaneous' integral formulas available, the antidifferentiation process quickly turns into a pattern-matching quest once the formal calculus course is passed.

6.3.1) Five Basic Antidifferentiation Rules

The five basic antidifferentiation rules stated in term of traditional integral/integration formulas are given below without proof. If a rule completes the antidifferentiation process, as in **R1**, **R3**, and **R4**, then the necessary constant C is traditionally shown. If a rule states an antidifferentiation process improvement, as in **R2** and **R5**, the constant C is not shown. But, the constant C must be shown once the antidifferentiation process is completed.

Three comprehensive examples are given at the end of the set.

R1. Antiderivative of a Constant: $\int k\,dx = kx + C$

$$\text{Case for } k = 1: \int dx = x + C$$

The special case for the value $k = 1$ states that the antiderivative for a differential is the variable itself (to within an arbitrary constant). This leads to the following *fundamental equality stream relating differentials, derivatives, and antidifferentiation.*

$$\frac{dy}{dx} = f(x) \Rightarrow dy = f(x)dx \Rightarrow$$

$$\int dy = \int f(x)dx \Rightarrow y = \int f(x)dx + C$$

R2: Coefficient Rule: If k is a constant, $\int kf(x)dx = k\int f(x)dx$

R3. Power Rule:

$$\int x^n dx = \frac{x^{n+1}}{n+1} + C, n \neq -1$$

R3. Power Rule (continued):

$$\int x^{-1} dx = \int \frac{1}{x} dx = \ln|x| + C, n = -1$$

R4: Exponential Rule Base e: $\int e^x dx = e^x + c$

R5. Sum/Difference Rule:

$$\int [f(x) \pm g(x)]dx = \int f(x)dx \pm \int g(x)dx$$

Ex 6.3.1: Find $F(x)$ for $f(x) = 8x^7 - \dfrac{8}{x^7} + 3e^x$.

$$F(x) = \int (8x^7 - 8x^{-7} + 3e^x)dx \Rightarrow$$

$$F(x) = \int 8x^7 dx - \int 8x^{-7} dx + \int 3e^x dx \Rightarrow$$

$$F(x) = 8\int x^7 dx - 8\int x^{-7} dx + 3\int e^x dx \Rightarrow$$

$$F(x) = 8\frac{x^8}{8} - 8\left[\frac{x^{-6}}{-6}\right] + 3e^x + C \Rightarrow$$

$$F(x) = x^8 + \frac{4}{3x^6} + 3e^x + C$$

*Note: As in differentiation, there are both process improvement rules (**R2&R5**) and process completion rules (**R1&R3&R4**). All five rules are meant to be used in concert with each other.*

Ex 6.3.2: Find $\int [9x^4 - 4x^3 + 6x + \dfrac{3}{x} + 2]dx$.

Here, neither $f(x)$ or $F(x)$ is explicitly stated. We proceed with

the understanding that $f(x) = 9x^4 - 4x^3 + 6x + \dfrac{3}{x} + 2$.

$$\int [9x^4 - 4x^3 + 6x + \frac{3}{x} + 2]dx =$$

$$\int 9x^4 dx - \int 4x^3 dx + \int \frac{3}{x} dx + \int 2dx =$$

$$9\int x^4 dx - 4\int x^3 dx + 3\int x^{-1} dx + 2\int dx =$$

$$9\frac{x^5}{5} - 4\frac{x^4}{4} + 3\ln |x| + 2x + C$$

Ex 6.3.3: Find $f(x)$ if $f'(x) = \dfrac{9}{\sqrt[3]{x}}$ and $f(1) = 1$

$\overset{1}{\mapsto} : f(x) = \int \left[\dfrac{9}{\sqrt[3]{x}} \right] dx + C \Rightarrow$

$f(x) = \int 9x^{-\frac{1}{3}} dx + C \Rightarrow$

$f(x) = 9 \int x^{-\frac{1}{3}} dx + C \Rightarrow$

$f(x) = \dfrac{27\sqrt[3]{x^2}}{2} + C$

****

$\overset{2}{\mapsto} : f(1) = 1 \Rightarrow$

$\dfrac{27\sqrt[3]{1^2}}{2} + C = 1 \Rightarrow$

$C = 1 - \dfrac{27}{2} = -\dfrac{25}{2} \Rightarrow$

$f(x) = \dfrac{27\sqrt[3]{x^2} - 25}{2}$

There are at least three acceptable ways to write the final answer as shown on the next page. The way we use is solely a matter of how we subsequently use the answer.

1. Emphasizing the antidifferentiation process

$\int [9x^4 - 4x^3 + 6x + \dfrac{3}{x} + 2] dx =$

$\dfrac{9x^5}{5} - \dfrac{4x^4}{4} + \dfrac{6x^2}{2} + 3\ln|x| + 2x + C$

2. Emphasizing the newly created antiderivative

$F(x) = \dfrac{9x^5}{5} - \dfrac{4x^4}{4} + \dfrac{6x^2}{2} + 3\ln|x| + 2x + C$

3. Emphasizing the original function/derivative relationship.

$$y = \frac{9x^5}{5} - \frac{4x^4}{4} + \frac{6x^2}{2} + 3\ln|x| + 2x + C \ \ or$$

$$f(x) = \frac{9x^5}{5} - \frac{4x^4}{4} + \frac{6x^2}{2} + 3\ln|x| + 2x + C$$

As the reader can well discern, context is everything when dealing with the symbols annotating the process and products associated with antidifferentiation.

6.3.2) Two Advanced Antidifferentiation Rules

R6. Parts Rule: $\int f(x)g'(x)dx = f(x)g(x) - \int g(x)f'(x)dx$

The Parts rule is the *reverse* of the product rule for differentiation and is proved using the product rule as a starting point.

Proof: Let $y = f(x) \cdot g(x)$

$$\frac{dy}{dx} = f(x)g'(x) + g(x)f'(x) \Rightarrow$$
$$dy = [f(x)g'(x) + g(x)f'(x)]dx$$

Antidifferentiating both sides of the last expression results in

$$\int dy = \int [f(x)g'(x) + g(x)f'(x)]dx \Rightarrow$$
$$y = \int f(x)g'(x)dx + \int g(x)f'(x)dx$$

Recalling that $y = f(x) \cdot g(x)$, we have

$$f(x) \cdot g(x) = \int f(x)g'(x)dx + \int g(x)f'(x)dx$$

Rearranging and neglecting the constant C, since the Parts rule is a process improvement rule, we obtain the desired result:

Integration by Parts—a Most Flexible Rule.

$$\int f(x)g'(x)dx = f(x)g(x) - \int g(x)f'(x)dx \therefore$$

The Parts rule is most useful when trying to find an antiderivative for an expression having a mixed nature (for example, an expression which is part algebraic and part transcendental). However, the Parts rule is somewhat tricky to use as success in finding an antiderivative depends on one's choice for $f(x)$ and $g'(x)$. When using the Parts rule, intuition guided by experience is the best approach. As with differentiation, *experience equals practice*. Notice that the Parts rule is a *staged rule*. This means that when we apply the Parts rule, we are still left with a remaining piece $\int g(x)f'(x)dx$ of the antidifferentiation process, a piece yet to be evaluated. Thus, the total process can be quite long and tedious.

Ex 6.3.4: Evaluate $\int x^2 e^x dx$

Stage 1: let $f(x) = x^2, g'(x) = e^x$

$$\int x^2 e^x dx = x^2 e^x - \int 2xe^x dx = x^2 e^x - 2\int xe^x dx$$

Stage 2: let $f(x) = x, g'(x) = e^x$ and evaluate $\int xe^x dx$

$$\int xe^x dx = xe^x - \int e^x dx = xe^x - e^x + C$$

By algebraically assembling the parts and pieces, we finally obtain

$$\int x^2 e^x \, dx = x^2 e^x - 2[xe^x - e^x + C] =$$

$$x^2 e^x - 2xe^x + 2e^x - 2C = (x^2 - 2x + 2)e^x + \hat{C}$$

where $\hat{C} = -2C$ and $\hat{C}$ remains an arbitrary constant.

Note: When arbitrary constants are algebraically combined with other numbers, the final algebraic expression is just as arbitrary. Hence, it is customary to give the whole expression the same name as the arbitrary constant embedded within it. Thus, in the above, we let $C \equiv \hat{C}$.

To verify the answer, simply take the derivative of the expression $(x^2 - 2x + 2)e^x + C$. If the antidifferentiation process is properly performed, then $[(x^2 - 2x + 2)e^x + C]'$ should match the original derivative $x^2 e^x$ (left to the reader).

Ex 6.3.5: Evaluate $\int x^2 \ln(x) dx$

Stage 1: let $f(x) = \ln(x), g'(x) = x^2$

$$\int x^2 \ln(x) dx = \frac{x^3}{3} \ln(x) - \int \frac{x^3}{3} \cdot \frac{1}{x} dx =$$

$$\frac{x^3}{3} \ln(x) - \frac{1}{3} \int x^2 dx = \frac{x^3}{3} \ln(x) - \frac{1}{9} x^3 + C$$

Since $\int \frac{x^3}{3} \cdot \frac{1}{x} dx$ can be directly evaluated, there is no need for a second stage.

You are once again encouraged to check by differentiating.

R7. Chain Rule: $\int f'(g(x))g'(x) dx = f(g(x)) + C$

The Chain rule is an obvious reverse of the Chain rule for differentiation and will not be proved.

Sometimes, the Chain rule is also written as $\int F'(g(x))g'(x)dx = F(g(x))+C$ in order to emphasize the antiderivative.

There are two often-used special cases of the Chain Rule (some people would say three) as shown below.

Special Case 1:

$$f'(x) = e^x \Rightarrow \int e^{g(x)}g'(x)dx = e^{g(x)} + C$$

Special Case 2: also known as the Generalized Power Rule

$$f'(x) = x^n \Rightarrow$$

$$\int [g(x)]^n g'(x)dx = \frac{[g(x)]^{n+1}}{n+1} + C, n \neq -1$$

$$\int [g(x)]^{-1} g'(x)dx = \int \frac{g'(x)}{g(x)}dx = \ln|g(x)| + C, n = -1$$

Ex 6.3.6: Evaluate $\int \left[\sqrt{x^2+1} \cdot 2x + \frac{2x}{x^2+1} \right] dx$

$$\int \left[\sqrt{x^2+1} \cdot 2x + \frac{2x}{x^2+1} \right] dx =$$

$$\int \sqrt{x^2+1} \cdot 2x dx + \int \left[\frac{2x}{x^2+1} \right] dx =$$

$$\int (x^2+1)^{\frac{1}{2}} \cdot 2x dx + \int (x^2+1)^{-1} \cdot 2x dx =$$

$$\frac{2\sqrt{(x^2+1)^3}}{3} + \ln(x^2+1) + C$$

Ex 6.3.7: Find $f(x)$ if $f'(x) = 2xe^{x^2} + x$ and $f(0) = 1$.

$\overset{1}{\mapsto}: f(x) = \int (2xe^{x^2} + x)dx + C \Rightarrow$

$f(x) = \int e^{x^2} \cdot 2x\, dx + \int x\, dx + C \Rightarrow$

$f(x) = e^{x^2} + \dfrac{x^2}{2} + C$

****

$\overset{2}{\mapsto}: f(0) = 1 \Rightarrow$

$e^{0^2} + \dfrac{0^2}{2} + C = 1 \Rightarrow$

$C = 0 \Rightarrow$

$f(x) = e^{x^2} + \dfrac{x^2}{2}$

6.3.3) The Seven Rules and Some 'Tricks of the Trade'

As one can imagine, there are hundreds of techniques, most of which are algebraic in nature that can be used to *soften up* the expression—*technically called an integrand*—within the integral sign. This *softening up* is for the purpose of preparing the integrand for subsequent antidifferentiation.

Note: tricks have a way of becoming techniques as your facility with algebra increases. All techniques are used in hopes of changing an obstinate expression into a new but algebraically equivalent expression matching known antidifferentiation rules. In this book, we will illustrate via several examples just three of these techniques and how they allow the use of our seven antidifferentiation rules.

Technique 1: *Modify the Integrand to Conform to the Chain Rule*

Suppose one has the task to evaluate $\int f'(g(x))h(x)dx$ where $g'(x) = kh(x)$ and k is a constant.

The integral expression $\int f'(g(x))h(x)dx$ can be easily adjusted to conform to the sought-after expression $\int f'(g(x))g'(x)dx$ by the following modification stream

$$\int f'(g(x))h(x)dx = \int f'(g(x)) \cdot 1 \cdot h(x)dx =$$

$$\int f'(g(x)) \cdot \left[\frac{1}{k}\right] \cdot [k] \cdot h(x)dx = \frac{1}{k}\int f'(g(x)) \cdot [k \cdot h(x)]dx =$$

$$\frac{1}{k}\int f'(g(x)) \cdot g'(x)dx = \frac{f(g(x))}{k} + C$$

The final result states that if we *miss* the derivative $g'(x)$ by a constant k, then the *basic antiderivative pattern for the chain rule* is divided by the same k. In practice, it is better to employ this technique on a case-by-case basis, instead of remembering an additional formula, as shown in the next two examples.

Ex 6.3.8: Evaluate $\int (7x^2 + 1)^{15} xdx$.

Technique 1 is directly applicable as shown.

$$\int (7x^2 + 1)^{15} \cdot xdx = \int (7x^2 + 1)^{15} \cdot \left[\frac{1}{14}\right] \cdot 14 \cdot xdx =$$

$$\int (7x^2 + 1)^{15} \cdot \left[\frac{1}{14}\right] \cdot 14xdx = \frac{1}{14}\int (7x^2 + 1)^{15} \cdot 14xdx =$$

$$\left[\frac{1}{14}\right] \cdot \frac{(7x^2 + 1)^{15}}{15} + C = \frac{(7x^2 + 1)^{15}}{210} + C$$

To check, just differentiate (left to reader).

Ex 6.3.9: Evaluate $\int e^{7x} dx$.

$$\int e^{7x} dx = \int e^{7x} \cdot \left[\frac{1}{7}\right] \cdot 7 \cdot dx =$$

$$\frac{1}{7} \int e^{7x} \cdot 7 dx = \frac{e^{7x}}{7} + C$$

Ex 6.3.10: Evaluate $\int \left[\dfrac{e^{2x}}{(e^x)^2 + 4}\right] dx$

This example may look impossible at first glance, but a little algebraic pre-softening serves well. Also notice the deft interplay of logarithmic/exponential integral formulas along with the use of a standard log rule (Appendix C).

$$\int \left[\frac{e^{2x}}{(e^x)^2 + 4}\right] dx = \int \left[\frac{e^{2x}}{e^{2x} + 4}\right] dx =$$

$$\int \left[\frac{e^{2x} \cdot \left[\frac{1}{2}\right] \cdot 2}{e^{2x} + 4}\right] dx = \frac{1}{2} \int \frac{e^{2x} \cdot 2 dx}{e^{2x} + 4} =$$

$$\frac{1}{2} \ln(e^{2x} + 4) + C = \ln \sqrt{(e^{x^2} + 4)} + C$$

Ex 6.3.11: Evaluate $\int \dfrac{[\ln(y^2)]^3}{y} dy$

In this example, an adjustment by a constant is not necessary even though it may seem so at first thought. Always let the process unfold before taking any unnecessary action.

$$\int \frac{[\ln(y^2)]^3}{y} dy = \int \frac{[2\ln(y)]^3}{y} dy =$$

$$8 \int [\ln(y)]^3 \cdot \frac{1}{y} \cdot dy = 8 \cdot \frac{[\ln(y)]^4}{4} + C = 2[\ln(y)]^4 + C$$

We finish our discussion of this technique with a very stern warning about a common integration error.

Stern Warning

Never adjust $\int f(g(x))h(x)dx$ by a variable as in the

False Equality: $\int f(g(x))h(x)dx = \dfrac{1}{x} \int f(g(x)) \cdot x \cdot h(x)dx$

Only constants can be moved through the integral sign
as stated by antidifferentiation rule **R2**:

If k is a constant, then $\int kf(x)dx = k \int f(x)dx$

Hence, the integration attempt

$$\int (x^2+1)^3 dx \neq \frac{1}{2x} \int (x^2+1)^3 \cdot 2x dx = \frac{(x^2+1)^4}{8x} + C$$

Is Totally Wrong!

Note: Believe it or not, I have seen all levels of mathematics students fall into the trap referred to by the warning. Don't do the same.

Technique 2: Use Ordinary Algebra to Simplify the Integrand

Ex 6.3.12: Evaluate $\int (x^2+1)^3 dx$.

So how do we evaluate $\int (x^2 + 1)^3 dx$? Answer: just cube it.

$$\int (x^2 + 1)^3 dx = \int (x^6 + 3x^4 + 3x^2 + 1) dx =$$

$$\frac{x^7}{7} + \frac{3x^5}{5} + x^3 + x + C$$

Note: Multiplying or dividing out is sometimes the only way to soften up the integrand as we prepare it for the process of antidifferentiation/integration.

Ex 6.3.13: Evaluate $\int \dfrac{(x+2)^4}{x} dx$.

$$\int \frac{(x+2)^4}{x} dx =$$

$$\int \left[\frac{x^4 + 8x^3 + 24x^2 + 32x + 16}{x} \right] dx =$$

$$\int \left[x^3 + 8x^2 + 24x + 32 + \frac{16}{x} \right] dx =$$

$$\frac{x^4}{4} + \frac{8x^3}{3} + 12x^2 + 32x + 16 \ln|x| + C$$

Ex 6.3.14: Evaluate $\int (5x + \sqrt{x})^2 dx$.

In this example, one must remember both the algebraic rules for radicals and the rules for their exponent equivalents.

$$\int (5x + \sqrt{x})^2 dx =$$

$$\int (25x^2 + 10x\sqrt{x} + x)dx =$$

$$\int (25x^2 + 10x^{\frac{3}{2}} + x)dx =$$

$$\frac{25x^3}{3} + 10\frac{x^{\frac{5}{2}}}{\frac{5}{2}} + \frac{x^2}{2} + C =$$

$$\frac{25x^3}{3} + 4x^{\frac{5}{2}} + \frac{x^2}{2} + C$$

Technique 3: *Expand the Integrand Using Partial Fractions*

By creating a least common denominator, we can add $\frac{1}{1+x}$ to $\frac{1}{1-x}$, obtaining the equality

$$\frac{1}{1+x} + \frac{1}{1-x} = \frac{2}{1-x^2}.$$

The antiprocess for this addition process is the splitting of the final result $\frac{2}{1-x^2}$ back into the component fractions

$$\frac{2}{1-x^2} = \frac{1}{1+x} + \frac{1}{1-x}.$$

The method used for performing the antiprocess is known as the technique of partial fractions. We will not develop the technique of partial fractions in this book, but only illustrate its use via our last two examples in this section. Applicable partial fraction formulas are in **Appendix A**.

Ex 6.3.15: Evaluate $\int \left[\frac{2}{1-x^2} \right] dx$.

$$\int \left[\frac{2}{1-x^2} \right] dx =$$

$$\int \left[\frac{1}{1+x} + \frac{1}{1-x} \right] dx =$$

$$\int \left[\frac{1}{1+x} \right] dx + \int \left[\frac{1}{1-x} \right] dx =$$

$$\int \left[\frac{1}{1+x} \right] dx - \int \left[\frac{-1}{1-x} \right] dx =$$

$$\ln|1+x| - \ln|1-x| + C =$$

$$\ln\left| \frac{1+x}{1-x} \right| + C$$

Ex 6.3.16: Evaluate $\int \left[\dfrac{x^2 + x + 1}{x^2(x+1)} \right] dx$.

$$\int \left[\frac{x^2 + x + 1}{x^2(x+1)} \right] dx =$$

$$\int \left[\frac{1}{x^2} + \frac{1}{x+1} \right] dx =$$

$$\int \left[\frac{1}{x^2} \right] dx + \int \left[\frac{1}{x+1} \right] dx =$$

$$\int x^{-2} dx + \int \left[\frac{1}{x+1} \right] dx =$$

$$\frac{x^{-1}}{-1} + \ln|x+1| + C =$$

$$\ln|x+1| - \frac{1}{x} + C$$

204

Ex 6.3.17: Evaluate $\int\left[\dfrac{e^x+x}{xe^x}\right]dx$.

$$\int\left[\dfrac{e^x+x}{xe^x}\right]dx=$$

$$\int\left[\dfrac{1}{x}+\dfrac{1}{e^x}\right]dx=$$

$$\int\dfrac{dx}{x}+\int e^{-x}dx=$$

$$\int\dfrac{dx}{x}+(-1)\int e^{-x}\cdot(-1)\cdot dx$$

$$\ln(x)-e^{-x}dx+C$$

$$\int_a^b \overset{\cdot\cdot}{\cup}dx$$

Section Exercises

1) Find $h(t)$ such that $h'(t)=t\sqrt{4-t^2}$ and $h(2)=2$.

2) Find $G(s)$ such that $g(s)=s^3 e^s$ and $G(0)=1$.

3) Find a polynomial $P(x)$ such that $P(x)$ has critical points at

$x=1$ and $x=-5$ and such that $P(1)=4$.

4) What is wrong with the following equality stream?

$$\int(x^4+1)^4 dx=\dfrac{1}{4x^3}\int(x^4+1)^4 4x^3 dx=\dfrac{(x^4+1)^5}{20x^3}+C$$

5) As indicated by the following integral expressions, antidifferentiate (*or integrate*) the associated integrands. Then check your results by differentiating the antiderivative.

a) $\int \sqrt[3]{x} dx$

b) $\int (x^3 + 3x^2 - x) dx$

c) $\int \left[x^3 + 3x^2 - \dfrac{1}{x^2} \right] dx$

d) $\int 20x(x^3 + 1)^2 dx$

e) $\int (\sqrt{x} + 1)^2 dx$

f) $\int (5x + 10) dx$

g) $\int (x + x^2 + x^3 + x^4 + x^5) dx$

i) $\int \ln(t) dt$

j) $\int \left[\dfrac{(x+2)(x^2 - 1)}{x - 1} \right] dx$

k) $\int e^x (e^x + 1)^2 e^{2x} dx$

l) $\int \left[\dfrac{(1 + \sqrt{x})^4}{\sqrt{x}} \right] dx$

m) $\int (2a^2 + 3)^{1001} \, ada$

n) $\int \dfrac{e^{\frac{10}{x}}}{x^2} dx$

o) $\int 2w(w^3 + 1)^{12} \, wdw$

6.4) Antidifferentiation Applied to Differential Equations

Ordinary *differential equations* equate two algebraic expressions where each the term(s) in each expression combine a *single independent variable*, the associated dependent variable, and associated derivatives of various orders. Four illustrations of differential equations are

1) $y'y + 2x^2 = x + y$ 2) $2y'' - 3y' + 4y = 0$

3) $2(yy')^2 = t + y''y$ 4) $y' = xy$

To solve a differential equation means to find a functional relationship between the dependent and independent variable that identically satisfies the given equation for all values of the independent variable. Such a function would have the general form $y = f(x)$ in illustrations 1) and 4) and the general form $y = f(t)$ in illustration 3). Either general form is acceptable in illustration 2) since the independent variable does not appear explicitly by name.

Note: Naming an implicit independent variable is generally the user's choice if there is no mentioned context such as time, where the independent variable is traditionally denoted by t .

In illustration 4) the function $y = f(x) = 4e^{\frac{x^2}{2}}$ is a solution to $y' = xy$ since $\left[4e^{\frac{x^2}{2}} \right]' \equiv x \cdot 4e^{\frac{x^2}{2}}$ for all x in the domain of f .

Each of the above differential equations has an *implicit-style formulation* meaning the derivative is not explicitly stated in terms of the independent variable as done in the differential equation $y' = 2t + 4$. As you might guess, explicitly stated differential equations are much easier to solve than implicitly stated differential equations. Both types of differential equations are used extensively to formulate many of the physical principles governing the universe.

For the serious student of calculus, the following statement should be somewhat eye-opening and sobering.

Note: partial differential equations, the term introduced below, are differential equations where several independent variables come into play. They are ever so briefly described in Chapter 9 and still await you in a formal course.

> Most modern applications of calculus encountered in science or engineering require the use of either *ordinary* or *partial* differential equations.

A simple but fundamental example, found just about everywhere in totally diverse contexts, is Isaac Newton's Second Law of Motion

$$F = (my')'$$

In words, Newton's Second Law states that the force applied to a body is equal to the *time-rate-of-change* of momentum of that very same body. The expression *time-rate-of-change* refers to a derivative where the independent variable is t. Also notice that momentum in itself is a product of mass and a first derivative, *velocity* in a *time-rate-of-change* context. *Thus, any problem involving force applied to a physical body and any subsequent motion experienced by the same body will require the use of a differential equation in order to formulate the appropriate mathematical model governing the body's motion.* Depending on the physical context, the actual solving of a particular differential equation *of motion* formulated via Newton's Second Law may be quite simple or extremely hard. Whether simple or hard, the solving of differential equations is one of the major tasks facing modern scientists and/or engineers. Because of their great flexibility in being able to model physical phenomena involving both changing and related variables, differential equations are here to stay. *And please ask yourself, what have you seen in this world or beyond that is not subject to change?*

Since differential equations stand at the very center of the calculus as it is practiced today, we will use differential equations to formulate all primary calculus applications remaining in this book.

This includes the use in Chapter 7 of a relatively simple differential equation to formulate the Second Fundamental Problem of Calculus—the problem of finding planar area—first presented in Chapter 3. But, before we can apply differential equations to actual problems, we must first learn how to solve them. Understand that the subject of differential equations is quite extensive requiring advanced courses over and above the basic calculus sequence found in colleges today. In this book, we only scratch the surface.

So, besides Newton's Law, where might a differential equation come from? A purely mathematical illustration will help clarify this question. Let's start with the function $y = \sqrt[3]{x^4 + 27}$. The following chain applies

$$y = \sqrt[3]{x^4 + 27} \Rightarrow$$
$$y^3 = x^4 + 27 \Rightarrow$$
$$[y^3]' = [x^4 + 27]' \Rightarrow$$
$$3y^2 y' = 4x^3 \Rightarrow$$
$$3y^2 y' - 4x^3 = 0$$

By definition, the end product $3y^2 y' - 4x^3 = 0$ is a differential equation. In this illustration, the differential equation has been generated through the process of implicit differentiation studied in Section 5.5.

A second question becomes, how do we recreate the function $y = \sqrt[3]{x^4 + 27}$, an obvious solution to $3y^2 y' - 4x^3 = 0$, from the information contained therein? The answer is: use the implicit-differentiation process in reverse. First, rewrite y' as a differential change ratio, i.e. $y' = \dfrac{dy}{dx}$. Next, substitute $y' = \dfrac{dy}{dx}$ into the differential equation $3y^2 y' - 4x^3 = 0$ to obtain

$$3y^2 \frac{dy}{dx} - 4x^3 = 0.$$

The last equality exposes the embedded differentials for both the independent and dependent variables, making a *true* differential equation as opposed to a derivative equation.

Note: the name differential equation comes from both the solution method used to solve differential equations (i.e. differentials) and the method by which these equations are primarily formulated (by differentials—more in Chapter 7).

Next, separate the independent and dependent variables

$$3y^2 dy = 4x^3 dx \ ,$$

a result which strongly suggests that we antidifferentiate both sides. A new but totally reasonable equality law (revisit Section 5.5) allows us to do just that.

Integral Law of Equality:

If $A(y)dy = B(x)dx$, then $\int A(y)dy = \int B(x)dx + C$.

With the Integral Law of Equality in hand, we can complete the process of recreating the original function $y = \sqrt[3]{x^4 + 27}$ from the separated differential equation $3y^2 dy = 4x^3 dx$ via the process of antidifferentiation:

$$3y^2 dy = 4x^3 dx \Rightarrow$$
$$\int 3y^2 dy = \int 4x^3 dx \Rightarrow$$
$$y^3 = x^4 + C \Rightarrow$$
$$y = \sqrt[3]{x^4 + C}$$

We have slightly missed the mark. Instead of recreating the original function $y = \sqrt[3]{x^4 + 27}$, we have recreated a more generic version $y = \sqrt[3]{x^4 + C}$.

So what additional information do we need in order to obtain recreate the exact function $y = \sqrt[3]{x^4 + 27}$? Answer: a boundary condition, in this case $y(0) = 3$, that allows the determination of C. Applying the boundary condition to $y = \sqrt[3]{x^4 + C}$ leads to an exact value for C, and, in turn, allows recreation of the original function $y = y(x)$.

$$y(x) = \sqrt[3]{x^4 + C} \Rightarrow$$
$$y(0) = \sqrt[3]{0^4 + C} = 3 \Rightarrow$$
$$\sqrt[3]{C} = 3 \Rightarrow C = 27 \Rightarrow$$
$$y(x) = \sqrt[3]{x^4 + 27}$$

Ex 6.4.1: Solve $y' = xy$ where $y(0) = 2$

The separation-of-variables technique will be the technique of choice. This technique usually entails two steps. Step 1) is the creation of a generic family of functions (via antidifferentiation) where each function in the family satisfies the differential equation without the boundary condition. Step 2) is the creation of a specific function that satisfies both the differential equation and boundary condition. This specific function is the final answer.

$$\overset{1}{\mapsto} : y' = xy \Rightarrow \frac{dy}{dx} = xy \Rightarrow$$
$$\frac{dy}{y} = xdx \Rightarrow$$
$$\int \frac{dy}{y} = \int xdx \Rightarrow$$
$$\ln|y| = \frac{x^2}{2} + C \Rightarrow$$
$$y = e^{\frac{x^2}{2} + C}$$

$$\stackrel{2}{\mapsto} : y(0) = 2 \Rightarrow$$

$$y(0) = e^{\frac{0^2}{2}+C} = 2 \Rightarrow$$

$$e^C = 2 \Rightarrow$$

$$y(x) = 2e^{\frac{x^2}{2}}$$

The separation-of-variables technique is only one of many solution techniques encountered when embarking on a serious study of ordinary differential equations. Nonetheless, the separation-of-variables technique is foundational for many of the advanced techniques presented in a later course. In this book, the separation-of-variables technique has sufficient enough power to solve all remaining differential equation formulations within the scope of this study. Let's review the solution process in general.

We must start with a differential equation that allows for— after replacement of y' with $\dfrac{dy}{dx}$ or $\dfrac{dy}{dt}$, etc.—algebraic separation of the dependent and independent variables, along with their respective differentials, across the equality sign. When successfully completed, this algebraic separation appears as follows:

$$g(y)dy = f(x)dx.$$

Note: It doesn't take much algebraic complexity for a differential equation to become unsolvable by the separation-of-variables technique. Raising derivatives to powers and having several orders of derivatives within the same equation can quickly render it unsolvable by the separation-of-variables technique.

The next step is the antidifferentiation of both sides of the equality.

$$\int g(y)dy = \int f(x)dx + C$$

By the Integral Law of Equality, this is a process which preserves the equality.

If $G(y)$ and $F(x)$ are two associated antiderivatives, then

$$G(y) = F(x) + C$$

We can use both sides of the above equality as input to $G^{-1}(y)$ in order to obtain the generic form of the sought-after function:

$$G^{-1}(G(y)) = G^{-1}(F(x) + C) \Rightarrow$$
$$y = y(x) = G^{-1}(F(x) + C)$$

Finally, applying a boundary condition, usually expressed as $y(a) = b$, locks in the arbitrary constant C by means of the equation

$$y(a) = G^{-1}(F(a) + C) = b \Rightarrow$$
$$G^{-1}(F(a) + C) = b$$

which is algebraically solved for C on a case-by-case basis.

Ex 6.4.2: Solve $\dfrac{y'}{x+1} = x^2$ where $y(1) = 5$

We will follow the same two-step process as in **Ex 6.4.1**.

$$\overset{1}{\mapsto} : \frac{y'}{x+1} = x^2 \Rightarrow y' = (x+1)x^2 \Rightarrow$$

$$\frac{dy}{dx} = (x+1)x^2 \Rightarrow$$

$$dy = (x+1)x^2 dx \Rightarrow$$

$$\int dy = \int (x+1)x^2 dx + C = \int [x^3 + x^2] dx + C \Rightarrow$$

$$y = y(x) = \frac{x^4}{4} + \frac{x^3}{3} + C$$

$$\overset{2}{\mapsto}: y(1) = 5 \Rightarrow$$

$$y(1) = \frac{1^4}{4} + \frac{1^3}{3} + C = 5 \Rightarrow$$

$$\frac{1^4}{4} + \frac{1^3}{3} + C = 5 \Rightarrow C = \frac{53}{12} \Rightarrow$$

$$y(x) = \frac{x^4}{4} + \frac{x^3}{3} + \frac{53}{12}$$

The above answer can be promptly confirmed by 1) checking the boundary condition for correctness and by 2) taking the derivative y' and substituting the algebraic expression obtained for y' into the differential equation $\dfrac{y'}{x+1} = x^2$ in order to see if an identity is, in fact, produced.

Ex 6.4.3: Solve $\dfrac{y'}{(x-2)^2} = y^2$ where $y(0) = 1$

$$\overset{1}{\mapsto}: \frac{y'}{(x-2)^2} = y^2 \Rightarrow \frac{y'}{y^2} = (x-2)^2 \Rightarrow$$

$$\frac{1}{y^2} \cdot \frac{dy}{dx} = (x-2)^2 \Rightarrow \frac{dy}{y^2} = (x-2)^2\, dx \Rightarrow$$

$$\int \frac{dy}{y^2} = \int (x-2)^2\, dx \Rightarrow$$

$$\frac{-1}{y} = \frac{(x-2)^3}{3} + C = \frac{(x-2)^3 + 3C}{3} \Rightarrow$$

$$y = y(x) = \frac{-3}{(x-2)^3 + C}$$

Note: see note comments on Page 196 in regard to manipulation and renaming of arbitrary constants. The same applies here.

****

$$\xmapsto{2}: y(0) = 1 \Rightarrow$$

$$y(0) = \frac{-3}{(0-2)^3 + C} = 1 \Rightarrow$$

$$\frac{-3}{-8+C} = 1 \Rightarrow C = 5 \Rightarrow$$

$$y(x) = \frac{-3}{(x-2)^3 + 5}$$

Ex 6.4.4: Newton's Second Law: An extremely large rivet is accidentally dropped from the top of the Sears Tower in Chicago. How many seconds does it take for the rivet to impact the street below? What is the velocity of the rivet at impact?

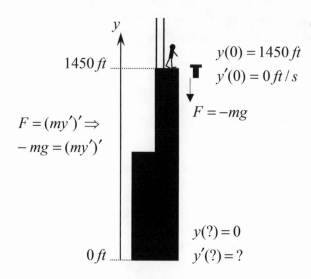

Figure 6.5: Newton, Sears, and the Rivet

Figure 6.5 is a sketch of the problem. The Sears Tower is $1450\,ft$ in height, and the twin antennas extend about $300\,ft$ further. Since the problem states that rivet is dropped from the roof, we will use $1450\,ft$ as the location for this incident.

Let $y = y(t)$ be the vertical position of the rivet in feet as a function of time in seconds. At $t = 0$ we will assume, having no better information than *accidentally dropped*, that the rivet is dropped precisely from the roofline $y(0) = 1450\,ft$ and that the initial velocity is zero $y'(0) = 0$. Once released, the only force acting on the rivet is the downward force due to the earth's gravity. This force is given by the expression $F = -mg$ where m is the mass of the rivet and $g = 32.2\frac{ft}{s^2}$ is the acceleration due to gravity near the earth's surface. The negative sign applies since the force due to gravity is acting opposite of increasing y. Applying Newton's Second Law gives the equality $-mg = (my')'$. Since the mass of the rivet is constant during its fall, the above equation reduces to the differential equation

$$-g = (y')' \Rightarrow -g = \frac{dy'}{dt} \Rightarrow \frac{dy'}{dt} = -g$$

where the last equality starts the formal solution process.

The solution process is a three-step approach. Step 1 is the construction (or reconstruction) of the velocity function from the given information. Step 2 is the construction of position as a function of time using the velocity function constructed in Step 1 as a starting point. Step 3 is the answering of the two questions using the two functions from Steps 1 and 2.

$$\overset{1}{\mapsto}: \frac{dy'}{dt} = -g \Rightarrow$$

$$dy' = -gdt \Rightarrow$$

$$\int dy' = \int -gdt + C \Rightarrow$$

$$y' = y'(t) = -gt + C \mapsto$$

$$y'(0) = 0 \Rightarrow -g \cdot 0 + C = 0 \Rightarrow$$

$$C = 0 \Rightarrow$$

$$y'(t) = -gt$$

****

$$\overset{2}{\mapsto}: y' = -gt \Rightarrow$$

$$\frac{dy}{dt} = -gt \Rightarrow$$

$$dy = -gtdt \Rightarrow$$

$$\int dy = \int -gtdt + C \Rightarrow$$

$$y = y(t) = -\tfrac{1}{2}gt^2 + C \mapsto$$

$$y(0) = 1450 \Rightarrow -\tfrac{1}{2}g \cdot 0^2 + C = 1450 \Rightarrow$$

$$C = 1450 \Rightarrow$$

$$y(t) = 1450 - \tfrac{1}{2}gt^2$$

$$\overset{3}{\mapsto}: \quad \text{To obtain the time at impact, set } y(t) = 0 \text{ (the position at impact) and solve for the corresponding time.}$$

$$0 = 1450 - \tfrac{1}{2}gt^2 \Rightarrow$$

$$16.1t^2 = 1450 \Rightarrow$$

$$t = 9.49\,\text{sec}$$

To obtain the velocity at impact, evaluate the velocity function obtained in Step 1 at $9.49 \sec$.

$$y'(9.49) = -305.58 \tfrac{ft}{s}$$

This is our first exposure to the tremendous power of differential equations which will be seen many times throughout the remainder of the book. In this example, we were able to construct a complete physical description of motion-versus-time (position, velocity, and acceleration) for a falling object via the use of antidifferentiation and Newton's Second Law as it is stated in terms of a differential equation.

We close this chapter with one more example. Our example is an engineering application that comes from the theory of heat and mass transfer where the governing physical principle is known as *Newton's Law of Cooling*.

Newton's Law of Cooling states that for an object of uniform temperature, the *time-rate* of object cooling is directly proportional to the temperature difference between the object and the surrounding medium. If $T(t)$ is the temperature of the object at a given time t, then Newton's Law of Cooling can be expressed as

$$T' = p(T - T_\infty)$$

where p is the constant of proportionality and T_∞ is the temperature of the surrounding medium. When using Newton's Law of Cooling, we must insure that the object is small enough (or has a large enough *thermal conductivity*) so that the uniform temperature distribution assumption is reasonable.

Ex 6.4.5: A small iron sphere is suspended from a ceiling by a piece of high-strength fishing line as shown in Figure 6.5. The air in the room below has a uniform temperature of $60^0 F$. It takes the ball one hour to cool from an initial temperature of $200^0 F$ to a temperature of $130^0 F$. How long does it take the ball to reach a temperature of $80^0 F$?

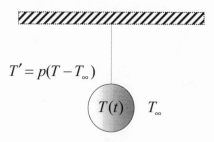

$$T' = p(T - T_\infty)$$

$$T(t) \quad T_\infty$$

Figure 6.6: Newton Cools a Sphere

The governing differential equation for this particular problem is

$$T' = \frac{dT}{dt} = p(T - 60), \text{ where } T(0) = 200 \text{ and } T(1) = 130.$$

Here, we have two known boundary conditions. Since there are two unknown constants, the integration constant C and the constant of proportionality p, both boundary conditions must be utilized in order to fix values for C and p.

$\overset{1}{\longmapsto}$: Solve the differential equation.

$$\frac{dT}{dt} = p(T - 60) \Rightarrow$$

$$dT = p(T - 60)dt \Rightarrow$$

$$\frac{dT}{t - 60} = pdt \Rightarrow$$

$$\int \frac{dT}{t - 60} = \int pdt + C \Rightarrow$$

$$\ln(T - 60) = pt + C \Rightarrow$$

$$T - 60 = e^{pt+C} \Rightarrow$$

$$T = T(t) = 60 + e^{pt+C}$$

$\overset{2}{\mapsto}$: Apply the two boundary conditions in order to determine C and p.

$T(0) = 200 \Rightarrow$

$60 + e^{p \cdot 0 + C} = 200 \Rightarrow$

$e^{C} = 140 \Rightarrow$

$T(t) = 60 + 140e^{pt}$

$T(1) = 130 \Rightarrow$

$60 + 140e^{p \cdot 1} = 130 \Rightarrow$

$e^{p} = .5 \Rightarrow p = \ln(.5) = -.693 \Rightarrow$

$T(t) = 60 + 140e^{-.693t}$

Conclusion: the object cools according to $T(t) = 60 + 140e^{-.693t}$.

We are now in a position to answer the original question. This is done by setting $T(t) = 80$ and solving for t.

$80 = 60 + 140e^{-.693t} \Rightarrow$

$e^{-.693t} = .14286 \Rightarrow$

$\ln(e^{-.693t}) = \ln(.14286) \Rightarrow$

$-.693t = -1.9459 \Rightarrow$

$t = 2.81hr$

$$\int_{a}^{b} \overset{..}{\cup} dx \quad \int_{a}^{b} \overset{..}{\cup} dx$$

Chapter Exercises

1) Solve the following differential equations and check.

a) $\dfrac{y'}{y^4} = xy + y : y(0) = 2$ b) $2yy' = \dfrac{2x+1}{x(x+1)} : y(1) = 1$

2) Suppose the rivet in **Ex 6.4** is tossed straight up at an initial velocity of $50 \frac{ft}{s}$ and an initial elevation of $1450\,ft$. On its way down, the rivet barely clears the roof of the Sears Tower and drops to the street below. Find the flowing four quantities: time to impact, impact velocity, highest elevation above street level, and time to achieve the highest elevation above street level.

3) A person having a body temperature of $98.6^0 F$ falls into an icy cold lake with water temperature $34^0 F$. 30 minutes later, severe hypothermia sets in when the person's average body temperature becomes $87^0 F$. The accident happens at 10:00PM and rescuers finally arrive on the scene at 1:00A.M. If the human body cools below $65^0 F$, survival is doubtful. Using Newton's Law of cooling as a crude, but in this case a necessary, body-temperature estimation technique, what do you think the person's chances are for survival?

7) Solving the Second Problem

"We cannot in any manner glorify the
Lord and Creator of the universe than that in all things—
How small so ever they appear to our naked eyes,
But which have yet received the gift of life and power of increase—
We contemplate the display of his omnificence and perfections
With utmost gratification."   Anton van Leeuwenhoek

7.1) The Differential Equation of Planar Area

We are now ready to solve the Second Fundamental Problem of Calculus as stated in Chapter 3.

> Find the exact area for a planar region where
> at least part of the boundary is a general curve.

Let f be a function with $f(x) > 0$ and $f'(x)$ existing for all x on an interval $[a,b]$. The Second Problem translates to finding the exact numerical area of the shaded region as shown in **Figure 7.1**.

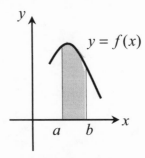

Figure 7.1: Planar Area with One Curved Boundary

The underlying premise that the shaded region has an exact numerical area is easily seen with the following thought experiment. Suppose we could construct a flat-sided beaker having a flat rectangular base.

Make the beaker so that two opposite parallel walls conform to the exact shape of the shaded region and let the inside distance between these walls be one unit. Let the top be closed except for one small nipple at the high point for the pouring in of water as shown in **Figure 7.2**. The right extension has been left in order to provide a handle. The left extension has been removed.

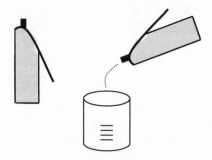

Figure 7.2: A Beaker Full of Area

If we fill our beaker of water, then the volume of the water is numerically equal to the shaded area since the inside distance between the two shaded walls is one unit. To measure the volume in the odd-shaped beaker, simply pour the water into a standard glass graduated container as again shown in **Figure 7.2**. If one follows this scenario, then the measured volume will be numerically equivalent to the area under the curve.

Continuing with the Second Problem, define the function $A(z)$ to be the area of the shaded region defined on the interval $[a, z]$ as shown in **Figure 7.3**.

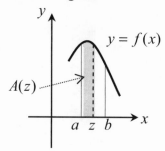

Figure 7.3: The Area Function

As a function, $A(z)$ associates a unique area for each $z \in [a,b]$. A common-sense examination of $A(z)$ will convince the reader that the following three properties hold:

1. $A(a) = 0$
2. $A(z)$ is increasing on $[a,b]$
3. $A(b)$ is the total area sought in **Figure 7.1**

We will now examine the differential behavior of $A(z)$. But first, an expression for dA in terms of z and dz must be created. The differential increment of area dA created when z is incremented by the differential dz is shown as the solid black line in **Figure 7.4**. This line is greatly magnified in the lower half of the figure, clearly showing the linear behavior (Chapter 5) of $y = f(x)$ on the differential interval $[z, z + dz]$.

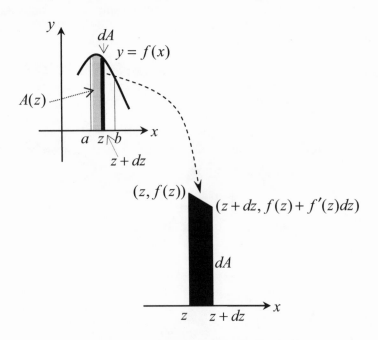

Figure 7.4: The Differential Increment of Area

The actual physical area of dA is the area of the black trapezoid and is given by (formula in Appendix B)

$$dA = \frac{1}{2}[f(z) + \{f(z) + f'(z)dz\}]dz \Rightarrow$$
$$dA = \frac{1}{2}[2f(z) + f'(z)dz\}]dz \Rightarrow$$
$$dA = f(z)dz + \frac{1}{2}f'(z)\{dz\}^2$$

The term $f(z)dz$ represents the area of a rectangle of height $f(z)$ and base dz. The term $\frac{1}{2}f'(z)\{dz\}^2$ represents the area of a triangle that is either added to or subtracted from the area of the rectangle depending on the sign of $f'(x)$. In **Figure 7.4**, it is subtracted. Whether added or subtracted, the term $\frac{1}{2}f'(z)\{dz\}^2$ doesn't affect the magnitude dA since it is second order and can be ignored per the rules of differentials stated in Chapter 4. Ignoring the second-order term leads to the fundamental differential equality highlighted below.

Note: we made sure that $f'(x)$ would not give us an unbounded-size problem in the term $\frac{1}{2}f'(z)\{dz\}^2$ by requiring the existence of $f'(x)$ for all x in the interval $[a,b]$.

The Differential Equation of Planar Area

$$dA = f(z)dz$$

The beauty and simplicity of this result is astounding. In words it states that the differential increment of area dA at z is given by the area of an incredibly thin rectangle—a mere sliver—having a height of $f(z)$ and a width of dz. Not only is this result dimensionally correct, it also makes perfect common sense. *Note: It has always filled me with a sense of wonder.*

Let the wonders continue! The differential equation $dA = f(z)dz$ also has a boundary condition given by $A(a) = 0$.

Together, the two pieces of information comprise a completely solvable differential-equation system, as we will now demonstrate.

Solve: $dA = f(z)dz, A(a) = 0$

Let $F(z)$ be any antiderivative for $f(z)$. Then

$$dA = f(z)dz \Rightarrow$$
$$\int dA = \int f(z)dz + C \Rightarrow$$
$$A = A(z) = F(z) + C \mapsto$$
$$A(a) = 0 \Rightarrow$$
$$F(a) + C = 0 \Rightarrow$$
$$C = -F(a) \Rightarrow$$
$$A(z) = F(z) - F(a)$$

In particular, we would like to evaluate the total area under consideration, the area given by $A(b)$:

$$A(b) = F(b) - F(a)$$

Here, we introduce the process symbol $F(z)\big|_a^b \equiv F(b) - F(a)$, which allows our final result to be stated as follows:

Let f be a differentiable function with $f(x) > 0$ on an interval $[a, b]$ and let $F(x)$ be an antiderivative for f. Then the area $A(b)$ above the x axis and below the graph of $y = f(x)$ and between the two vertical lines $x = a$ and $x = b$ is given by

$$A(b) = F(z)\big|_a^b = F(b) - F(a)$$

The above equality is without a doubt the most famous result in elementary calculus. For this reason, it is rightfully called the **Fundamental Theorem of Calculus**.

What makes the previous result so fundamental is that identical differential techniques are used to solve both the First and Second Fundamental Problems of Calculus. In Chapter 7, our starting point for solving the area problem is the simple explicit differential equation $dA = f(z)dz$. This very same equation could have been just as easily used in Chapter 5 to find the derivative of $A(z)$ via one straightforward division: $\dfrac{dA}{dz} = f(z)$. Hence, *area problems and slope problems are two sides of the same coin.*

The result $A(b) = F(z)\big|_a^b = F(b) - F(a)$ will now be used to verify the area of a trapezoid which is perhaps the most complicated of the elementary polygonal figures.

Ex 7.1.1: Use the Fundamental Theorem of Calculus to find the area of the trapezoid defined by the x axis, the lines $x = 2$ and $x = 5$, and having upper boundary $f(x) = 2x + 3$.

Figure 7.5 shows the desired area with value $A = 30$ obtained via the use of the elementary trapezoid formula.　　We need to see if we can match this result using $A = F(z)\big|_a^b = F(b) - F(a)$ where the b in the interval $[a, b]$ is to be understood by the problem context.

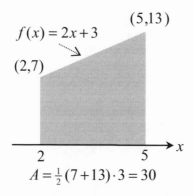

$$A = \tfrac{1}{2}(7 + 13) \cdot 3 = 30$$

Figure 7.5: Trapezoid Problem

An antiderivative for $f(x) = 2x + 3$ is $F(x) = x^2 + 3x$. Applying the Fundamental Theorem gives

$$A = x^2 + 3x \big|_2^5 =$$
$$(5^2 + 3 \cdot 5) - (2^2 + 3 \cdot 2) =$$
$$40 - 10 = 30$$

And, as advertised, the final result matches what we already know.

We end Section 7.1 with a classic example that allows us a first taste of computational mastery in the wondrous new world of curvilinear area.

Ex 7.1.2: Use the Fundamental Theorem of Calculus to find the area of the region defined by the x axis, the lines $x = 4$ and $x = 6$, and having upper boundary $f(x) = x^2 - 3x - 4$.

Figure 7.6 shows the desired area.

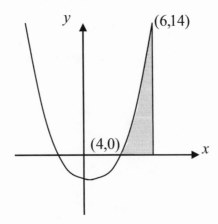

Figure 7.6: Area Under $f(x) = x^2 - 3x - 4$ on $[4,6]$

Note: in practice, a necessary first step in solving any area problem is the sketching of the area so we know exactly what area is to be evaluated. Many errors in area evaluation could be avoided if students would just sketch the area.

228

The reader may recognize the quadratic function $f(x) = x^2 - 3x - 4$ from Chapter 4. Since $f(x) \geq 0$ on $[4,6]$, the Fundamental Theorem can be used to evaluate the area by way of the antiderivative $F(x) = \dfrac{x^3}{3} - \dfrac{3x^2}{2} - 4x$. Hence

$$A = \left[\frac{x^3}{3} - \frac{3x^2}{2} - 4x \right] \Big|_4^6 =$$

$$\left[\frac{6^3}{3} - \frac{3 \cdot 6^2}{2} - 4 \cdot 6 \right] - \left[\frac{4^3}{3} - \frac{3 \cdot 4^2}{2} - 4 \cdot 4 \right] =$$

$$\left[\frac{216}{3} - \frac{108}{2} - 24 \right] - \left[\frac{64}{3} - \frac{48}{2} - 16 \right] =$$

$$\frac{152}{3} - \frac{60}{2} - 8 = \frac{152}{3} - 38 = \frac{152}{3} - \frac{114}{3} = \frac{38}{3}$$

A good computational practice is to keep denominators segregated in denominator-alike groups as long as possible. Combining unlike denominators early in the calculation usually means that you must do it twice, for terms in $F(b)$ and for terms in $F(a)$, doubling the chance for error.

$$\int_a^b \ddot\cup \, dx$$

Section Exercises

1. Find the area under the curve $f(x) = x^4 + x^2$ on a) the interval $[0,4]$, on b) the interval $[-4,4]$.

2. Find $z > 3$ so that the area under the curve $f(x) = x^2 + 2$ on the interval $[3, z]$ is equal to 10.

3. Use the Fundamental Theorem of Calculus to develop the general area formula for a trapezoid (Appendix B).

7.2) Process and Products: Continuous Sums

In **Figure 7.4** the differential increment of area dA is given by the expression $dA = f(z)dz$ once second order effects have been eliminated per the rules for differentials. This expression represents the area of an infinitesimally thin rectangle having height $f(z)$ and width dz. Suppose we were to start at $x = a$ and mark off successive increments of this same width dz until we finally stop at $x = a$. Since dz is a differential, it will take millions upon millions of these 'itsy bitsy' infinitesimal increments laid end to end, with no overlap, in order to traverse the total distance from $x = a$ to $x = b$. **Figure 7.5** shows just one increment of width dz as it occupies an infinitesimal fraction of the interval $[a, b]$. Likewise, the associated differential area dA occupies an infinitesimal fraction of the total area under the curve from $x = a$ to $x = b$.

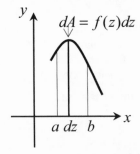

Figure 7.7: One 'Itsy Bitsy' Infinitesimal Sliver

Let's sum all the infinitesimal areas from $x = a$ to $x = b$. The symbol used to denote this one infinitesimal at-a-time summing process is

$$\sum_{x=a}^{x=b} dA = \sum_{x=a}^{x=b} f(z)dz .$$

How do we evaluate $\sum_{x=a}^{x=b} f(z)dz$? Once seen, both the evaluation process and the final answer are going to absolutely amaze you.

230

Let $F(z)$ be an antiderivative for $f(z)$. In picking dz for subsequent summation, we are going to make sure it is small enough so $F(z)$ has the standard differential linear behavior on each infinitesimal subinterval $[z, z + dz]$, i.e.

$$F(z + dz) = F(z) + f(z)dz \Rightarrow$$
$$F(z + dz) - F(z) = f(z)dz$$

Also, choose dz so that the right endpoint of the very last subinterval corresponds exactly with $x = b$. If all of this has been prearranged to be the case, then

$$\sum_{x=a}^{x=b} f(z)dz =$$

$$\sum_{x=a}^{x=b} [F(z + dz) - F(z)] =$$

$$[F(a + dz) - F(a)] +$$
$$[F(a + 2dz) - F(a + dz)] +$$
$$[F(a + 3dz) - F(a + 2dz)] +$$
$$[F(a + 4dz) - F(a + 3dz)] +$$
$$\dots +$$
$$[F(a + \{?\} \cdot dz) - F(a + \{?-1\} \cdot dz)] +$$
$$[F(b) - F(a + \{?\} \cdot dz)] =$$
$$F(b) - F(a)$$

As you can see, millions upon millions of tiny terms—the question mark signifies that the exact number is unknown—cancel in accordion fashion. This leaves us with just the two already-familiar macro terms $F(b)$ and $F(a)$. Hence,

$$\sum_{x=a}^{x=b} dA = \sum_{x=a}^{x=b} f(z)dz = F(b) - F(a).$$

Recall by Section 7.1, the total area A also equals $F(b) - F(a)$.

Hence, we have $A = \sum_{x=a}^{x=b} f(z)dz = F(b) - F(a)$.

Examining $A = \sum_{x=a}^{x=b} f(z)dz = F(b) - F(a)$ immediately suggests a known process for finding the sum: simply find an antiderivative $F(z)$ for $f(z)$ and then evaluate the expression $F(z)\big|_a^b$.

In Chapter 6, the integration sign $\int f(z)dz$ is the process symbol used to denote that an antiderivative is to be found. When finding a differential sum, the above integration sign is slightly modified as follows

$$\int_a^b f(z)dz.$$

With definition

$$\int_a^b f(z)dz \equiv F(z)\big|_a^b = F(b) - F(a) : F'(z) = f(z)$$

Note: the reader should show that the using of $F(z) + C$ for an antiderivative in $(F(z) + C)\big|_a^b$ the above expression is unnecessary due to the cancellation of the constant C.

The amazing chained expression

$$A = \int_a^b f(z)dz = F(z)\big|_a^b = F(b) - F(a) : F'(z) = f(z)$$

is quite suggestive in itself. The integration sign $\int$ looks like a smoothed-out version of the summation symbol $\sum$.

Indeed, that is exactly how the differential summing process works: by smoothly and continuously building up the whole from millions upon millions of tiny pieces $f(z)dz$. In this introductory example, the whole is a total area, but doesn't necessarily need to be as we shall soon discover. The numbers a and b signify the start and the end of the *continuous summing process*. Finally, the expression $F(z)\big|_a^b = F(b) - F(a)$ gives an alternative means, by way of antidifferentiation, for performing the *continuous summation*—a significant process improvement.

Note: In nature, large structures are also built from a vast number of tiny pieces: human bodies are built from cells and stars are built from atoms.

$$\int_a^b \smile dx$$

Section Exercises: None

7.3) Process Improvement: Definite Integrals

Let's take a brief moment and review the process and associated products for the following two expressions:

$$\int f(x)dx \text{ and } \int_a^b f(x)dx.$$

Recall that the expression $\int f(x)dx$ tells the user to find a family of antiderivatives for the differential $f(x)dx$. If $F(x)$ is one such family member with $F'(x) = f(x)$, then all other such family members can be characterized by $F(x) + C$ where C is an arbitrary constant. We then have that

$$\int f(x)dx = F(x) + C.$$

Note: $F(x)$ *without the constant* C *is sometimes called the fundamental or basic antiderivative.*

The process on the left-hand side is known either as indefinite integration (older terminology) or antidifferentiation (newer terminology). The associated product on the right-hand side, in this case a function, is known either as an indefinite integral or antiderivative. The term *indefinite* refers to the arbitrary constant C. Whether called indefinite integration or antidifferentiation, the symbol $\int$ is always known as an integral or an integration sign.

The symbol $\int_a^b f(x)dx$ carries the process of antidifferentiation (or indefinite integration) a step further than required by $\int f(x)dx$. Not only is an antiderivative $F(x)$ found, but additionally, the antiderivative is evaluated at the two endpoints a and b per the process-to-product evaluation scheme

$$\int_a^b f(x)dx = F(x)\,|_a^b = F(b) - F(a).$$

The process in this case is called definite integration and the associated product is called a definite integral. The integration process is viewed as definite since 1) the arbitrary constant C is no longer part of the final product and 2) the final product has a precise numerical value, which is very definite indeed.

Note: Somehow, the modern antidifferentiation/antiderivative terminology never came to use as a way of describing definite integration.

The symbol $\int_a^b f(x)dx$ is perhaps the most holographic in all of calculus. It can be interpreted in at least three different ways.

1. As a processing symbol for functions, $\int_a^b f(x)dx$ instructs the *operator* to start the process by finding the basic antiderivative $F(x)$ for $f(x)dx$ and finish it by evaluating the quantity $F(x)\,|_a^b = F(b) - F(a)$. This interpretation is pure process-to-product with no context.

234

2. As a summation symbol for differential quantities, $\int_a^b f(x)dx$ signals to the *operator* that myriads of infinitesimal quantities of the form $f(x)dx$ are being continuously summed on the interval $[a,b]$ with the summation process starting at $x = a$ and ending at $x = b$. Depending on the context for a given problem, such as area, the differential quantities $f(x)dx$ and subsequent total can take on a variety of meanings. This makes continuous summing a powerful tool for solving real-world problems as will be shown in subsequent sections. The fact that continuous sums can also be evaluated by $\int_a^b f(x)dx = F(x)\big|_a^b = F(b) - F(a)$ is a *fortunate consequence of the Fundamental Theorem of Calculus.*

3. Lastly, $\int_a^b f(x)dx$ can be interpreted as a point solution $y(b)$ to any explicit differential equation having the general form $dy = f(x)dx : y(a) = 0$, such as the differential equation of planar area discussed in Section 7.1. In this interpretation $\int_a^b f(x)dx$ is first modified by integrating over the arbitrary subinterval $[a, z] \subset [a,b]$ which results in the expression $y(z) = \int_a^z f(x)dx = F(z) - F(a)$. Substituting $x = a$ gives the stated boundary condition $y(a) = F(a) - F(a) = 0$ and substituting $x = b$ gives $y(b) = F(b) - F(a) = \int_a^b f(x)dx$. In this context, the function $y(z) = F(z) - F(a)$, as a unique solution to $dy = f(x)dx : y(a) = 0$, can also be interpreted as a continuous running sum from $x = a$ to $x = z$.

For the rest of Section 7.3, we will concentrate on interpreting the definite integral $\int_a^b f(x)dx$ as a computational processing formula using a rule that we will call the Rosetta Stone of Calculus. This is one of several possible alternate formulations of the Fundamental Theorem of Calculus.

The Rosetta Stone of Calculus

$$\int_a^b f(x)dx = F(x)\,|_a^b = F(b) - F(a) : F'(x) = f(x)$$

Like the original Rosetta Stone that allowed for decoding of Egyptian hieroglyphics, the Rosetta Stone of Calculus will allow for the swift and easy evaluation (a decoding if you will) of the definite integral $\int_a^b f(x)dx$. The amazing thing is that the evaluation is always the same irregardless of the contextual interpretation under which $\int_a^b f(x)dx$ was formulated, a major process improvement.

Ex 7.3.1: Evaluate the definite integral $\int_4^7 (2x+3)dx$.

$$\int_4^7 (2x+3)dx = (x^2 + 3x)\,|_4^7 =$$

$$(7^2 + 3\cdot 7) - (4^2 + 3\cdot 4) =$$

$$70 - 28 = 42$$

Note: As stated, this example has no context. The reader is encouraged to give it a context by letting $f(x) = 2x + 3$ be the upper bounding curve for a trapezoid defined on the interval $[4,7]$. In the context of planar area, is the answer reasonable?

Ex 7.3.2: Evaluate the definite integral $\int\limits_{1}^{4}\dfrac{xdx}{(x^2+3)}$ and interpret as a planar area.

$$\int\limits_{1}^{4}\dfrac{xdx}{(x^2+3)}=\dfrac{1}{2}\int\limits_{1}^{4}\dfrac{2xdx}{(x^2+3)}=$$

$$\dfrac{1}{2}\ln(x^2+3)\Big|_{1}^{4}=$$

$$\dfrac{1}{2}\ln(17)-\dfrac{1}{2}\ln(4)=\dfrac{1}{2}\ln(4.25)$$

Since $f(x)=\dfrac{x}{x^2+3}>0$ for all $x\in[1,4]$, $\int\limits_{1}^{4}\dfrac{xdx}{(x^2+3)}$ can be interpreted as the area between the x axis and $f(x)$ from $x=1$ to $x=4$.

Note: In this example, the first step is the obtaining of the antiderivative using all known rules and methodologies discussed in Chapter 6. Only after the antiderivative is obtained, do we substitute the two numbers 4 and 1. These two numbers are formally called the upper and lower limits of integration—yet another deviation from the title of this book.

Ex 7.3.3: Evaluate the definite integral $\int\limits_{2}^{5}(7x^3-4x^2+x+1)dx$.

$$\int\limits_{2}^{5}(7x^3-4x^2+x+1)dx=$$

$$\left(\dfrac{7x^4}{4}-\dfrac{4x^3}{3}+\dfrac{x^2}{2}+x\right)\Big|_{2}^{5}=$$

$$\left(\dfrac{7\cdot5^4}{4}-\dfrac{4\cdot5^3}{3}+\dfrac{5^2}{2}+5\right)-\left(\dfrac{7\cdot2^4}{4}-\dfrac{4\cdot2^3}{3}+\dfrac{2^2}{2}+2\right)=$$

$$\left(\frac{4375}{4}-\frac{500}{3}+\frac{25}{2}+5\right)-\left(\frac{112}{4}-\frac{32}{3}+\frac{4}{2}+2\right)=$$

$$\left(\frac{4375}{4}-\frac{112}{4}\right)-\left(\frac{500}{3}-\frac{32}{3}\right)+\left(\frac{25}{2}-\frac{4}{2}\right)+(5-2)=$$

$$\frac{4263}{4}-\frac{468}{3}+\frac{21}{2}+3=$$

$$\frac{12789}{12}-\frac{1872}{12}+\frac{126}{12}+\frac{36}{12}=\frac{11079}{12}=\frac{3693}{4}$$

Ex 7.3.3 is typical of the stepwise precision necessary when evaluating the definite integral of a polynomial function. I have always cautioned my students to keep the inevitable rational terms in denominator-alike groups, adding and subtracting within a group as necessary. The final result from each group can then be converted to equivalent fractions having like denominators, preparing them for the grand total. By proceeding in this fashion, we utilize the error-prone like-denominator process only once.

$$\int_{a}^{b}\ddot{\smile}\,dx$$

Section Exercises

1) Evaluate the following definite integrals

a) $\displaystyle\int_{0}^{2}\sqrt{4x+1}\,dx$

b) $\displaystyle\int_{3}^{7}\left(x^{3}+x^{2}+3x+5\right)dx$

c) $\displaystyle\int_{e}^{4e}\frac{(\ln x)^{2}}{x}\,dx$

d) $\displaystyle\int_{0}^{1}(2x+1)^{10}\,dx$

2) Use a definite integral to find the total area above the x axis and below $y=x+\sqrt{x}$ on the interval $[1,4]$.

7.4) Geometric Applications of the Definite Integral

In this section, continuous sums are used in order to obtain planar areas, volumes of revolution, surface areas of revolution, and arc lengths. Each of these aforementioned quantities can also be formulated in terms of a differential equation. Whether framed in terms of a continuous sum or differential equation, the solution for the associated volume, or area, or surface area, or arc length will be given in terms of a definite integral, which is indistinguishable in either case.

7.4.1) Planar Area Between Two Curves

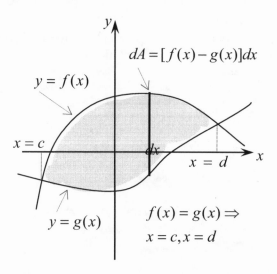

Figure 7.8: Area Between Two Curves

Suppose we were required to find the area of the shaded region shown in **Figure 7.8**. A differential element of rectangular area would take the form $dA = [f(x) - g(x)]dx$ where the function $f(x)$ is the upper bounding curve, and the function $g(x)$, the lower bounding curve.

The expression $[f(x)-g(x)]dx$ is always positive—no matter where the overall figure happens to be located within the four quadrants—if one remembers that $f(x)$ is to be the upper bounding curve (in the direction of increasing y), and $g(x)$ is to be the lower bounding curve as shown **in Figure 7.8**. Hence, all differential areas of the general form $dA=[f(x)-g(x)]dx$ are positive. This definitely needs to be the case if we are trying to sum millions upon millions of tiny quantities in order to make a total area.

To obtain the shaded area, simply use continuous summing to add up all the differential elements of area having general form $dA=[f(x)-g(x)]dx$. Start the summing process at $x=c$ and finish the process at $x=d$, where the two endpoints $c\,\&\,d$ can readily be obtained by setting $f(x)=g(x)$ and solving for x. The general setup for finding the area between two curves can be expressed in terms of the definite integral

$$A = \int_{c}^{d}[f(x)-g(x)]dx$$

or the differential equation

$$dA =[f(x)-g(x)]dx : A(c)=0.$$

Per previous discussion, solving the differential equation and finding the particular value $A(d)$ also corresponds to evaluating

$$\int_{c}^{d}[f(x)-g(x)]dx,$$

again making the definite integral our magnificent tool of choice.

Ex 7.4.1: Find the area between the two curves $f(x) = 6-x^2$ and $g(x) = 3-2x$.

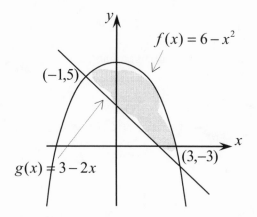

Figure 7.9: Area Between $f(x) = 6 - x^2$ **and** $g(x) = 3 - 2x$

Figure 7.9 shows the desired area. *Note: it is a necessity to draw the area before one evaluates the same. Only a drawing can allow us to determine the relative positions of the two curves. Relative position— upper boundary respect to lower boundary—is of prime importance when formulating the quantity* $dA = [f(x) - g(x)]dx$, *which could be rephrased in verbal terms as* $dA = [upper - lower]dx$.

First, we solve for the endpoints:

$$f(x) = g(x) \Rightarrow$$
$$6 - x^2 = 3 - 2x \Rightarrow$$
$$3 + 2x - x^2 = 0 \Rightarrow$$
$$(3 - x)(1 + x) = 0 \Rightarrow$$
$$x = 3 \,\&\, {-1}$$

Next, we set up the associated differential area.

$$dA = [f(x) - g(x)]dx =$$
$$[(6 - x^2) - (3 - 2x)]dx =$$
$$[3 + 2x - x^2]dx$$

241

Finally, we evaluate $\int_{-1}^{3}(3+2x-x^2)dx$, interpreting it as planar area obtained by a continuous summing process.

$$A = \int_{-1}^{3}(3+2x-x^2)dx =$$

$$\left[3x+x^2-\frac{x^3}{3}\right]\Big|_{-1}^{3} =$$

$$\left[9+9-\frac{27}{3}\right]-\left[-3+1+\frac{1}{3}\right] =$$

$$18-\frac{27}{3}+2-\frac{1}{3} = 20-\frac{28}{3} = \frac{32}{3}$$

Ex 7.4.2: Find the area between the curve $f(x)=x^2-3x-4$ and the x axis on the interval $[0,6]$.

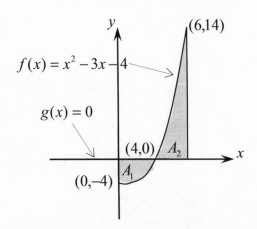

Figure 7.10: Over and Under Shaded Area

Figure 7.10 shows the desired area. Define $g(x) \equiv 0$.

242

The interval of interest $[0,6]$ has both endpoints given. Thus, we don't have to solve for endpoints in this example. However, we do have to solve for the crossover point since the upper and lower bounding curves reverse their relative positions.

$$f(x) = g(x) \Rightarrow$$
$$x^2 - 3x - 4 = 0 \Rightarrow$$
$$(x-4)(x+1) = 0 \Rightarrow$$
$$x = 4, \& -1$$

The value $x = -1$ is outside $[0,6]$, the interval of consideration. Thus, $x = 4$ marks the one crossover point. It is not much of a stretch to see that $A_{Total} = A_1 + A_2$. Since $f(x)$ and $g(x)$ switch roles on $[0,6]$, we must set up two separate definite integrals—with *upper* and *lower* properly placed in each—in order to evaluate the total area.

$$A_{Total} = A_1 + A_2 =$$
$$\int_0^4 [\{0\} - \{x^2 - 3x - 4\}]dx + \int_4^6 [\{x^2 - 3x - 4\} - \{0\}]dx =$$
$$\int_0^4 (-x^2 + 3x + 4)dx + \frac{38}{3} =$$
$$\frac{58}{3} + \frac{38}{3} = \frac{96}{3} = 32$$

7.4.2) Volumes of Revolution

Under suitable restrictions, the graph of a function $f(x)$ can be rotated about the x axis, or the y axis, or a line parallel to one axis, or the other. The rotation of the locus of points defined by the graph of $f(x)$ about a fixed axis sweeps out a surface area and an associated interior volume.

Both of these quantities, surface area of revolution and volume of revolution, can be ascertained by use of the definite integral. In this subsection, we will focus on determining *volume of revolution*.

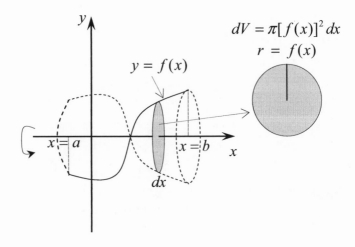

Figure 7.11: Volume of Revolution Using Disks

Figure 7.11 shows the graph of a function $f(x)$ being rotated counterclockwise about the x axis on the interval $[a,b]$. A goblet shape will be generated in this fashion where either end could serve as the end used for drinking. The stem will be a single point of zero thickness as shown on the graph. To determine the volume of the goblet, first remove a slice of thickness dx (the gray-shaded region), which has been cut orthogonal to the x axis (the axis of rotation). This differential slice has a circular cross section with frontal area given by $\pi[f(x)]^2$. Multiplying by the thickness dx gives the volume of the differential slice $dV = \pi[f(x)]^2 dx$. Hence, the total volume V of the goblet can be obtained quite easily by continuous summation of all the differential quantities dV starting at $x = a$ and ending at $x = b$.

Hence, the total volume is given in terms of the definite integral

$$V = \int_a^b \pi [f(x)]^2 \, dx \, .$$

The same integral is also point solution (at $x = b$) to the associated differential equation for the volume of revolution:

$$dV = \pi [f(x)]^2 \, dx : V(a) = 0 \, .$$

Again, both roads lead to the same definite integral no matter which thought process we use, continuous sum or differential equation, to interpret and subsequently solve the problem.

Ex 7.4.3: A) Use the method of disks to find the volume of rotation when the graph of $f(x) = x^2 - 1$ is revolved about the x axis on the interval $[0,2]$. B) Find the volume when $f(x)$ is revolved about the line $y = 3$ and C) about the line $y = -2$.

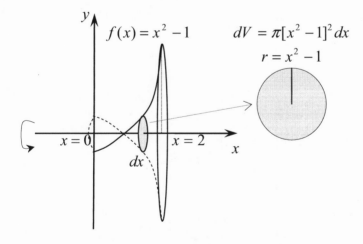

Figure 7.12: Rotating $f(x) = x^2 - 1$ about the x axis

Part A: **Figure 7.12** shows $f(x) = x^2 - 1$ as it is rotated around the x axis. Even though the functional values are negative on the subinterval $[01]$, dV is never negative due to the squaring of $f(x)$. *Note: One of the advantages of the disk method for finding a volume of revolution is that dV never turns negative during the continuous summing process. Hence, one doesn't have to adjust $f(x)$ on subintervals where $f(x) < 0$ in order to maintain positive dVs.*

The total volume is given by

$$V = \int_0^2 \pi[x^2 - 1]^2 \, dx =$$

$$\pi \int_0^2 [x^4 - 2x^2 + 1]dx = \pi\left[\frac{x^5}{5} - \frac{2x^3}{3} + x\right]\Big|_0^2 =$$

$$\pi\left[\frac{2^5}{5} - \frac{2 \cdot 2^3}{3} + 2\right] - [0] = \pi\left[\frac{32}{5} - \frac{16}{3} + 2\right] = \frac{46\pi}{15}$$

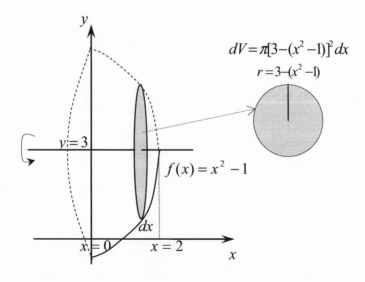

$$dV = \pi[3 - (x^2 - 1)]^2 \, dx$$

$$r = 3 - (x^2 - 1)$$

$y = 3$

$f(x) = x^2 - 1$

dx

$x = 0$ $x = 2$

Figure 7.13: Rotating $f(x) = x^2 - 1$ about the Line $y = 3$

Part B: As shown in **Figure 7.13**, $dV = [3-(x^2-1)]^2\, dx$ is the appropriate differential volume element. *You should verify why this is so.* Once dV is determined, you can determine the corresponding V by

$$V = \int_0^2 \pi[3-(x^2-1)]^2\, dx = \int_0^2 \pi[4-x^2]^2\, dx$$

$$\pi \int_0^2 [x^4 - 8x^2 + 16]dx = \pi \left[\frac{x^5}{5} - \frac{8x^3}{3} + 16x \right] \Big|_0^2 =$$

$$\pi \left[\frac{2^5}{5} - \frac{8 \cdot 2^3}{3} + 32 \right] - [0] = \pi \left[\frac{32}{5} - \frac{64}{3} + 32 \right] = \frac{256\pi}{15}$$

Note: Volume of revolution problems using the disk method can be quite tricky when rotating about an axis other than the x axis. Great care must be taken to draw a representative disk and associated frontal area. Of prime importance is the relationship of the function $f(x)$ to the radius of the disk so drawn. There is no golden rule except think it through on a case by case basis.

Part C: No figure is shown. The reader should verify that the differential element of volume dV is given by the expression $dV = [2+(x^2-1)]^2\, dx$. Hence

$$V = \int_0^2 \pi[2+(x^2-1)]^2\, dx = \int_0^2 \pi[1+x^2]^2\, dx$$

$$\pi \int_0^2 [x^4 + 2x^2 + 1]dx = \pi \left[\frac{x^5}{5} + \frac{2x^3}{3} + x \right] \Big|_0^2 =$$

$$\pi \left[\frac{2^5}{5} + \frac{2 \cdot 2^3}{3} + 2 \right] - [0] = \pi \left[\frac{32}{5} + \frac{16}{3} + 2 \right] = \frac{206\pi}{15}$$

Ex 7.4.4: Verify that the volume of a right circular cone is given by the expression $V = \frac{1}{3}\pi r^2 h$ where r is the radius of the base and h is the altitude.

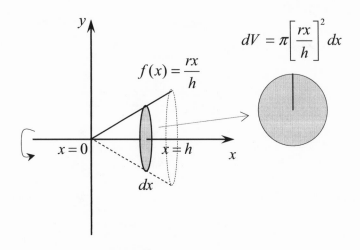

Figure 7.14: Verifying the Volume of a Cone

Figure 7.14 shows the setup for this problem. A right circular cone can be generated by revolving the line given by $f(x) = \dfrac{rx}{h}$ about the x axis on the interval $[0,h]$. The associated differential element of volume is given by $dV = \pi \left[\dfrac{rx}{h}\right]^2 dx$. The volume V is given by

$$V = \int\limits_0^h \pi \left[\frac{rx}{h}\right]^2 dx = \frac{\pi r^2}{h^2} \int\limits_0^h x^2 dx = \frac{\pi r^2}{h^2} \left[\frac{x^3}{3}\right] \Big|_0^h =$$

$$\frac{\pi r^2}{h^2}\left[\frac{h^3}{3}\right] = \frac{\pi r^2 h}{3} \quad \therefore$$

248

The disk method for finding a volume of revolution is the method of choice when rotating about the axis representing the independent variable (or a line parallel to this axis) as shown in **Figure 7.15**.

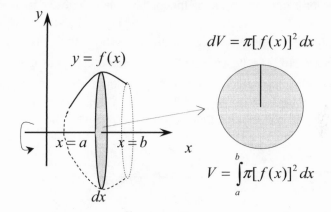

Figure 7.15: When to Use the Disk Method

But, suppose we wish to rotate this same graph about the y axis from $x = a$ to $x = b$. In this case, the y axis represents the dependent variable. Hence, the associated method of choice for finding a volume of revolution is the method of cylindrical shells as shown in **Figure 7.16**.

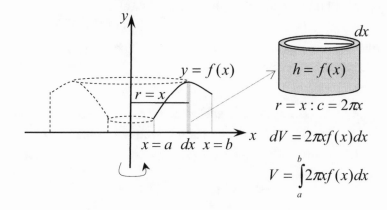

Figure 7.16: Method of Cylindrical Shells

249

A cylindrical shell is akin to a short piece of thin copper tubing (much like that used in home construction). When the graph of $f(x)$ is rotated about the y axis, it can be thought of as sweeping out millions upon millions of these thin cylindrical shells where each has thickness dx. At any particular x location in the interval $[a,b]$, the surface area of the shell is given by $A = 2\pi x f(x)$. Multiplying by the associated thickness gives the associated differential element of volume $dV = 2\pi x f(x)dx$. **Figure 7.17** shows the shell in **Figure 7.16** after it has been cut and flattened out, better exposing all three dimensions.

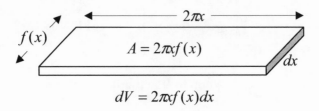

$$dV = 2\pi x f(x)dx$$

Figure 7.17: Flattened Out Cylindrical Shell

Unlike the disk method, when using cylindrical shells in a continuous summing process in order to build up a total volume, care must be taken to insure that $f(x)$ is always positive. Hence, in practice, $dV = 2\pi x \,|\, f(x)\,|\, dx$. Using $|\,f(x)\,|$ guarantees a positive A and associated dV which, in turn, guarantees no dV cancellation as we build up the total volume V.

Ex 7.4.5: A) Use cylindrical shells to find the volume of rotation when the graph of $f(x) = x^2 - 1$ is revolved about the y axis on the x interval $[0,2]$. B) Find the volume when $f(x)$ is revolved about the line $x = -3$.

Part A: **Figure 7.18** shows the volume of revolution to be determined. In this example the function $f(x) = x^2 - 1$ is negative on the subinterval $[0,1]$. Therefore, we will need to use $dV = 2\pi x \, | \, x^2 - 1 \, | \, dx$ as our differential volume element.

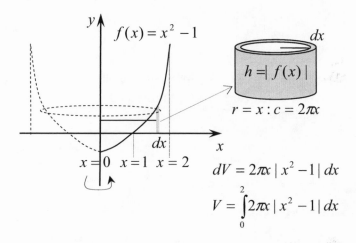

$$f(x) = x^2 - 1$$

$$dx$$

$$h = | f(x) |$$

$$r = x : c = 2\pi x$$

$$x = 0 \quad x = 1 \quad x = 2$$

$$dV = 2\pi x \, | \, x^2 - 1 \, | \, dx$$

$$V = \int_0^2 2\pi x \, | \, x^2 - 1 \, | \, dx$$

Figure 7.18: Rotating $f(x) = x^2 - 1$ about the y axis

The associated V is given by the definite integral

$$V = \int_0^2 2\pi x \, | \, x^2 - 1 \, | \, dx = \int_0^1 2\pi x (1 - x^2) dx + \int_1^2 2\pi x (x^2 - 1) dx \, ,$$

which must be split (as shown) into two integrals where the first sums the volume elements (per a properly-signed $f(x)$) on the interval $[0,1]$ and the second sums the volume elements on $[1,2]$. Continuing with the evaluation:

$$V = \int_0^1 2\pi x (1 - x^2) dx + \int_1^2 2\pi x (x^2 - 1) dx =$$

$$2\pi \left[\int_0^1 (x - x^3) dx + \int_1^2 (x^3 - x) dx \right] \Rightarrow$$

$$V = 2\pi\left[\left\{\frac{x^2}{2} - \frac{x^4}{4}\right\}\Big|_0^1 + \left\{\frac{x^4}{4} - \frac{x^2}{2}\right\}\Big|_1^2\right] =$$

$$2\pi\left[\left\{\frac{1^2}{2} - \frac{1^4}{4}\right\} - \{0\} + \left\{\frac{2^4}{4} - \frac{2^2}{2}\right\} - \left\{\frac{1^4}{4} - \frac{1^2}{2}\right\}\right] =$$

$$2\pi\left[\left\{\frac{1}{4}\right\} + \{2\} + \left\{\frac{1}{4}\right\}\right] = 5\pi$$

Part B: No figure is shown. The reader is to verify that the radius of rotation is now $x + 3$. Accordingly, the desired volume is given by the definite integral

$$V = \int_0^2 2\pi(x+3)\,|\,x^2 - 1\,|\,dx =$$

$$\int_0^1 2\pi(x+3)(1-x^2)dx + \int_1^2 2\pi(x+3)(x^2-1)dx =$$

$$2\pi\left[\int_0^1 (x - x^3)dx + \int_1^2 (x^3 - x)dx\right] +$$

$$6\pi\left[\int_0^1 (1 - x^2)dx + \int_1^2 (x^2 - 1)dx\right] =$$

$$5\pi + 6\pi\left[\left\{x - \frac{x^3}{3}\right\}\Big|_0^1 + \left\{\frac{x^3}{3} - x\right\}\Big|_1^2\right] =$$

$$5\pi + 6\pi\left[\left\{1 - \frac{1^3}{3}\right\} - \{0\} + \left\{\frac{2^3}{3} - 2\right\} - \left\{\frac{1^3}{3} - 1\right\}\right] =$$

$$5\pi + 6\pi\left[\left\{\frac{2}{3}\right\} + \left\{\frac{2}{3}\right\} + \left\{\frac{2}{3}\right\}\right] = 17\pi$$

Note: notice how we able to use the results from Part A in order to ease our workload in Part B.

Ex 7.4.6: Rotate the function $f(x) = \dfrac{rx}{h}$ in **Ex 7.4.4** about the y axis and show that the role of r and h is reversed in the conic volume formula when creating the cone as shown in **Figure 7.19**—i.e. volume in this case is given by the formula $V = \dfrac{\pi h^2 r}{3}$.

Figure 7.19 shows the function being rotated about the y axis. The cone we want to create is an inverted cone of radius h and height r.

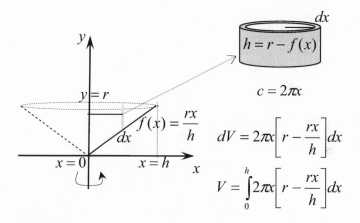

Figure 7.19: The Volume of an Inverted Cone

This example is a bit trickier than what it appears to be at first glance. The height of a cylindrical shell associated with the inverted cone is given by the expression $r - \dfrac{rx}{h}$, and not the expression $\dfrac{rx}{h}$. The associated differentiate volume element is $dV = 2\pi x\left[r - \dfrac{rx}{h}\right] dx$.

We are to continuously sum these differential volume elements on the interval $[0, h]$. Hence

$$V = \int_0^h 2\pi x \left[r - \frac{rx}{h} \right] dx =$$

$$2\pi \int_0^h \left[rx - \frac{rx^2}{h} \right] dx =$$

$$2\pi \left[\frac{rx^2}{2} - \frac{rx^3}{3h} \right] \Big|_0^h =$$

$$2\pi \left[\left\{ \frac{rh^2}{2} - \frac{rh^3}{3h} \right\} - \{0\} \right] =$$

$$2\pi \left[\frac{rh^2}{6} \right] = \frac{\pi h^2 r}{3} \therefore$$

Notice that the method of cylindrical shells again verifies a known result—another testimony to the power of calculus.

As a final comment, the disk method and cylindrical shell method offer a great deal of computational flexibility to the user provided the setup is correct. The errors most often made are setup errors and include: the wrong radius, the wrong height, the wrong limits of integration, or a combination thereof. The initial drawing of an accurate picture, as in any word problem, showing a properly chosen differential volume element is the key to success.

7.4.3) Arc Length

Our next topic addresses the issue of finding the arc length (or curve length) for a function $f(x)$ defined on an interval $[a, b]$. **Figure 7.20** shows arc length, traditionally denoted by the letter s, for such a function and the associated differential methodology used to obtain it.

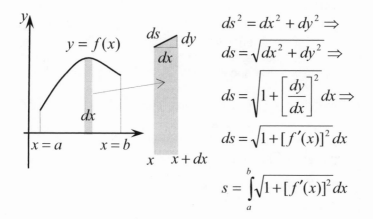

$$ds^2 = dx^2 + dy^2 \Rightarrow$$

$$ds = \sqrt{dx^2 + dy^2} \Rightarrow$$

$$ds = \sqrt{1 + \left[\frac{dy}{dx}\right]^2}\, dx \Rightarrow$$

$$ds = \sqrt{1 + [f'(x)]^2}\, dx$$

$$s = \int_a^b \sqrt{1 + [f'(x)]^2}\, dx$$

Figure 7.20: Arc Length and Associated Methodology

Since the function $f(x)$ has linear behavior on any differential interval $[x, x + dx]$, the associated differential element of arc length ds (shown by the dark line capping the trapezoid in **Figure 7.20**) is a straight line segment. By the Pythagorean Theorem, we have that $ds = \sqrt{dx^2 + dy^2}$. Since $dy = f'(x)dx$ on a differential interval $[x, x + dx]$, the differential element of arc length reduces to $ds = \sqrt{1 + [f'(x)]^2}\, dx$ after some algebraic manipulation. Continuous summing of these differential elements from $x = a$ to $x = b$ is accomplished by the definite integral

$$s = \int_a^b \sqrt{1 + [f'(x)]^2}\, dx$$

where s is the desired arc length.

Note: Little did Pythagoras realize where his theorem would eventually appear. Can you see the Pythagorean Theorem embodied in the tremendous result (framed in terms of a definite integral) above?

A good thing about the differential element of arc length ds is that $ds > 0$ no matter the sign of $f'(x)$. This means in practice that intervals don't have to be broken into subintervals in order to adjust for a negative differential element: as sometimes is the case in planar area between two curves or volumes of revolution using cylindrical shells. A frustrating thing about arc length is that $s = \int_a^b \sqrt{1+[f'(x)]^2}\, dx$ can become extremely hard to evaluate, even when $f(x)$ is quite simple. Techniques beyond the scope of this introductory volume must then be employed in order to obtain an antiderivative for $\sqrt{1+[f'(x)]^2}$.

Ex 7.4.7: Find the arc length of the graph of $f(x) = x^{\frac{3}{2}}$ on the interval $[0,1]$. The figure associated with arc length is always identical to the graph of $f(x)$ itself. In this case, we shall dispense with the figure since f is a relatively simple function to visualize and defined everywhere on $[0,1]$. Continuing

$$\overset{1}{\mapsto} : f'(x) = \tfrac{3}{2} x^{\frac{1}{2}} \Rightarrow$$

$$ds = \sqrt{1+[\tfrac{3}{2} x^{\frac{1}{2}}]^2}\, dx \Rightarrow ds = \frac{\sqrt{4+9x}}{2}\, dx$$

$$\overset{2}{\mapsto} : s = \int_0^1 \frac{\sqrt{4+9x}}{2}\, dx = \tfrac{1}{18} \int_0^1 (4+9x)^{\frac{1}{2}} \cdot 9 \cdot dx =$$

$$\frac{(4+9x)^{\frac{3}{2}}}{27} \Big|_0^1 = \frac{(13^{\frac{3}{2}}-4)}{27} \cong 1.5878$$

As an order-of-magnitude check, simply compute the straight line distance between the two endpoints of the graph $(0,0)$ and $(1,1)$. The answer is $\sqrt{2} = 1.414$, which is the shortest distance between two points; and, as it should, $1.5878 > \sqrt{2}$.

Ex 7.4.8: Find the arc length of the graph of $f(x) = x^2$ on the interval $[0,1]$. Again, no graph is shown due to the visual simplicity of the example.

$$\overset{1}{\mapsto}: f'(x) = 2x \Rightarrow$$
$$ds = \sqrt{1 + [2x]^2}\, dx \Rightarrow ds = \sqrt{1 + 4x^2}\, dx$$

$$\overset{2}{\mapsto}: s = \int_0^1 \sqrt{1 + 4x^2}\, dx$$

Now—believe it or not—we are stuck. and our function is quite simple. We have no means to evaluate the above integral utilizing the methods presented in this book.

And, again, the following is totally wrong:

$$s = \int_0^1 (1 + 4x^2)^{\frac{1}{2}}\, dx \neq \frac{1}{8x} \int_0^1 (1 + 4x^2)^{\frac{1}{2}} \cdot 8x \cdot dx$$

Functions known as trigonometric functions and inverse trigonometric functions are needed in order to construct an antiderivative for the expression $\sqrt{1 + 4x^2}$. So, in this example, I will just state the correct answer, which is 1.47815, and ask you to perform an order-of-magnitude check as done in **Ex 7.4.7**.

   We'll stop here with our arc length discussion. As stated at the beginning of this subsection, it doesn't take much of a function to create a definite integral that in today's vernacular is a "real bear". Even authors of standard "full-up" calculus texts carefully pick their examples (insuring that they are fully doable) when addressing this topic.

7.4.4) Surface Area of Revolution

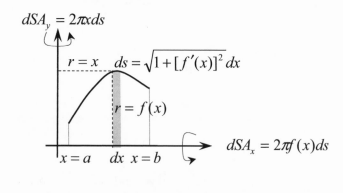

Figure 7.21: Two Surface Areas of Revolution Generated by One Graph

Figure 7.21 shows the differential element of arc length ds in **Figure 7.20** being rotated about the x axis. When ds is rotated in this fashion, it sweeps out an associated differential element of surface area given by $dSA_x = 2\pi x f(x)ds$. The total of all such elements from $x = a$ to $x = b$ can be found by evaluating the definite integral

$$SA_x = \int_a^b 2\pi f(x)\sqrt{1+[f'(x)]^2}\,dx$$

The quantity SA_x is called the surface area of revolution for the function f, rotated about the x axis from $x = a$ to $x = b$. Likewise, the same function f can be rotated about the y axis from $x = a$ to $x = b$. The associated surface area of revolution is given by

$$SA_y = \int_a^b 2\pi x\sqrt{1+[f'(x)]^2}\,dx$$

Since there is nothing that disallows $f(x)$ being negative on subintervals or the whole of $[a,b]$, the formula for SA_x should be modified (similar to what we did in the case for cylindrical shells) to read

$$SA_x = \int_a^b 2\pi \, | f(x) | \sqrt{1+[f'(x)]^2} \, dx \, .$$

With this change, we can proceed safely with the actual evaluation of surface areas of revolution, splitting the interval $[a,b]$ into subintervals as necessary to accommodate for the negativity of f.

Ex 7.4.9: Find SA_y for $f(x) = x^2$ on the interval $[0,2]$.

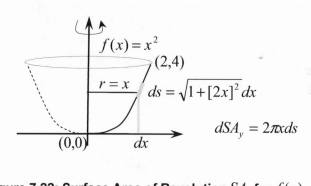

Figure 7.22: Surface Area of Revolution SA_y for $f(x) = x^2$

Figure 7.22 shows the desired surface area of revolution. Setting up the appropriate definite integral, we have

$$\overset{1}{\mapsto} : SA_y = \int_a^b 2\pi x \sqrt{1+[f'(x)]^2} \, dx \Rightarrow$$

$$SA_y = \int_0^2 2\pi x \sqrt{1+4x^2} \, dx$$

$$\mapsto: \int_0^2 2\pi x\sqrt{1+4x^2}\,dx =$$

$$\frac{2\pi}{8}\int_0^2 (1+4x^2)^{\frac{1}{2}}\cdot 8x\cdot dx =$$

$$\frac{\pi}{6}(1+4x^2)^{\frac{3}{2}}\,\Big|_0^2 =$$

$$\frac{\pi}{6}[(1+4\cdot 2^2)^{\frac{3}{2}}-(1+4\cdot 0^2)^{\frac{3}{2}}] =$$

$$\frac{\pi}{6}[(17)^{\frac{3}{2}}-(1)^{\frac{3}{2}}] \Rightarrow$$

$$SA_y \cong 36.1769$$

An important point needs to be made regarding this example. *Notice that it is workable using the methods in this book.* Rotating $f(x)$ about any other line of the form $x=-r$ parallel to the y axis leads to the following surface area integral

$$SA_{x=-r} = \int_0^2 2\pi(x+r)\sqrt{1+4x^2}\,dx =$$

$$\int_0^2 2\pi x\sqrt{1+4x^2}\,dx + \int_0^2 2\pi r\sqrt{1+4x^2}\,dx$$

The first definite integral is again workable. But alas, the second definite integral is now unworkable by elementary methods. Hence, as previously stated in the subsection addressing arc length, it doesn't take a great deal of algebraic change to turn a workable problem into an unworkable problem (at least by elementary methods). Additionally, if **Ex 7.4.9** had called for the calculation of SA_x, then appropriate definite integral, given

by $SA_x = \int_0^2 2\pi x^2\sqrt{1+4x^2}\,dx$, would have also been found to be unworkable by elementary methods.

Ex 7.4.10: Find SA_x for $f(x) = x^3$ on the interval $[0,1]$.

No graph is shown to the visual simplicity.

$$\overset{1}{\mapsto} : SA_x = \int_a^b 2\pi f(x)\sqrt{1+[f'(x)]^2}\,dx \Rightarrow$$

$$SA_x = \int_0^1 2\pi x^3 \sqrt{1+9x^4}\,dx$$

$$\overset{2}{\mapsto} : \int_0^1 2\pi x^3 \sqrt{1+9x^4}\,dx =$$

$$\frac{2\pi}{36}\int_0^1 (1+9x^4)^{\frac{1}{2}} \cdot 36x^3 \cdot dx =$$

$$\frac{\pi}{27}(1+9x^4)^{\frac{3}{2}}\,|_0^1 =$$

$$\frac{\pi}{27}[(1+9\cdot1^4)^{\frac{3}{2}} - (1+9\cdot0^4)^{\frac{3}{2}}] =$$

$$\frac{\pi}{27}[(10)^{\frac{3}{2}} - (1)^{\frac{3}{2}}] \Rightarrow$$

$$SA_y \cong 3.5631$$

In this example, $SA_y = \int_0^1 2\pi x\sqrt{1+9x^4}\,dx$, a definite integral which again requires advanced methods for completion.

$$\int_a^b \overset{\bullet\bullet}{\cup}\,dx \quad \int_a^b \overset{\bullet\bullet}{\cup}\,dx$$

Chapter/Section Exercises

Note: in the following exercises, part of the challenge is the identifying of an appropriate definite integral—planar area, volume of revolution, arc length, or surface area of revolution—to apply in the specific instance.

1. Use a definite integral to verify that the total surface area for a frustum (**Figure 7.23**) of height h, lower radius b, and upper radius a is given by the formula

$$SA = \pi(a^2 + b^2) + 2\pi\left[\frac{a+b}{2}\right]\sqrt{h^2 + (b-a)^2}.$$

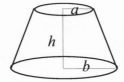

Figure 7.23: A Frustum

2. Use a definite integral to find the total volume of the frustum shown in **Figure 7.23**.

3. Use a definite integral to find the area between the two curves given by $f(x) = x^2 - 5x + 7$ and $g(x) = 4 - x$.

4. Use a definite integral to find the arc length for the graph of $f(x) = 7x + 1$ on the interval $[1,3]$. Verify by the distance formula.

5. Use $x^2 + y^2 = r^2$, the equation for a circle centered at the origin and with radius r, and appropriate definite integrals to verify that the surface area for a sphere is given by $SA = 4\pi r^2$ and the volume by $V = \frac{4}{3}\pi r^3$.

6. A) Find the area below the curve $y = x^3$ and above the x axis on the interval $[2,3]$. B) Find the volume of revolution generated when this same area is rotated about the x axis. C) The y axis. D) The line $y = -2$. E) The line $x = -2$. *Note: B, C, D & E are all workable.*

8) Sampling the Power of Differential Equations

"It suddenly struck me that
That tiny pea, pretty and blue, was the Earth.
I put up my thumb and shut one eye,
And my thumb blotted out the planet Earth.
I didn't feel like a giant; I felt very, very small."
Neil Armstrong

8.1) Differential Equalities

In one sense, Calculus can be thought of as the mathematical art of using the small to measure or build the large. Just as Neil Armstrong raised one tiny thumb to metaphorically examine the entirety of planet Earth, we can use the tiny differential to examine all sorts of physical and human phenomena on scales far exceeding what is capable via direct experience or observation.

Chapter 7 introduced us to several new differential equalities. Differential equalities are simply equations that relate two or more tiny differentials through some algebraic means. A primary example of a differential equality is $dy = f'(x)dx$, first derived in Chapter 4. Hence, a *differential equality* is nothing more than a *differential equation*. The expression $dy = f'(x)dx$ can then be thought of as the original differential equation, encountered quite early in our study.

The Original Differential Equation

$$dy = f'(x)dx$$

Consequently, far from being an exclusively advanced topic in calculus, we see that differential equations soon arise after the differential concept is developed as an immediate and natural follow-on. In **Table 8.1** below, we list the various differential equations encountered thus far with the associated application.

Chapter and Equation	Application
4: $dy = f'(x)dx$	Function Building
6: $F = (my')'$	2nd Law of Motion
6: $dT = p(T - T_\infty)dt$	Law of Cooling
7: $dA = \mid f(z) \mid dz$	Planar Area
7: $dA = [f(x) - g(x)]dx$	Planar Area between Curves
7: $dV = \pi[f(x)]^2 dx$	Volume of Revolution: Disks
7: $dV = 2\pi x \mid f(x) \mid dx$	Volume of Revolution: Shells
7: $ds = \sqrt{1 + [f'(x)]^2} dx$	Arc Length
7: $dSA_x = 2\pi \mid f(x) \mid \sqrt{1 + [f'(x)]^2} dx$	Surface Area of Revolution: x axis
7: $dSA_y = 2\pi x\sqrt{1 + [f'(x)]^2} dx$	Surface Area of Revolution: y axis

Table 8.1: Elementary Differential Equations

The two new differential equations introduced in Chapter 6 are from physics. Both equations are highly flexible and can be used to solve diverse problems. We will see more of Newton's 2nd Law in this chapter. What makes the equations in Chapter 7 unique is that all of them can be solved using the continuous summation interpretation of the definite integral. This interpretation is possible because any geometric quantity can be thought of as the summation of many tiny pieces.

Once this conceptualization is in place, the trick (or the artistry) reduces to the characterization of a generic differential piece by an appropriate differential equality. The sought-after whole is then assembled from *millions upon millions* of differential pieces via the continuous summing process, a process whose product can be conveniently generated using the definite integral.

In Chapter 8, we are going to expand our use of differential equations by exploring applications not exclusively geometric. Two major areas will be sampled, physics and finance. Both are very diverse and are diverse from each other. But, we shall soon marvel at the flexible power of problem formulation by the skilled use of differential equations, a power that allows the trained user to readily develop the *mathematical micro-blueprint* associated with a variety of phenomena in the heavens, on the earth, and in the marketplace. *And from the micro blueprint, we can then build a model for the associated macro phenomena.*

$$\int_a^b \overset{\bullet\bullet}{\cup} dx$$

Section Exercises

1. As a review of Section 6.4, solve the following differential equations and characterize as either *implicit* or *explicit*.

a) $xdy = y^2 x^3 dx : y(1) = 2$

b) $y' = x^2 + x : y(0) = 3$

c) $xy' = 2y' + \dfrac{1}{y} : y(4) = 1$

d) $y' = 4y : y(0) = 1$

2. Let $dA = kx^4 : A(2) = 2$ be a differential equation associated with planar area. Determine the constant k so that $A(4) = 16$. Evaluate $A(6)$.

8.2) Applications in Physics

8.2.1) Work, Energy, and Space Travel

Work is a topic typically found in a chapter on applications of the definite integral and is typically introduced right after surface areas, volumes of revolution, and arc lengths. In this book, we will use work as a *bridge topic*. In doing so, we will bounce back and forth between continuous summation interpretations and more fluid interpretations requiring alternative approaches to the solving of work associated differential equations. Eventually, we will bounce into space and escape from planet earth.

The classic definition of work is given by the algebraic expression $W = F \cdot D$. The force F is assumed to be constant and aligned in a direction parallel to the distance D through which the force is applied. As long as F is constant and aligned parallel to D, the above definition holds, and problems are somewhat easy to solve. *Note: When F and D are not aligned, then we need to employ the methods of vector analysis, which is way beyond the scope of this book. Hence, we shall stay aligned.*

To illustrate, suppose $F = 10 lb_f$ and acts in alignment through a distance $D = 7 ft$. Then the total work performed is $W = (10 lb_f) \cdot (7 ft)$ or $70 ft \cdot lb_f$ (read *foot-pounds*). **Figure 8.1** depicts a typical work situation as introduced in elementary physics texts: where a constant force F aligned with the x axis is being applied by the stick person in moving a box through a distance $D = b - a$.

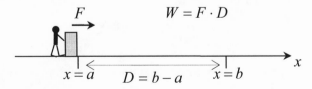

Figure 8.1: Classic Work with Constant Force

It doesn't take much modification to turn the classic work situation into one requiring the use of calculus in order to obtain a solution. All one has to do is make the applied force non-constant as the effort proceeds from $x = a$ to $x = b$. By doing so, F now becomes a function of the position variable x (i.e. $F = F(x)$); and work (as opposed to the basic macro-definition $W = F \cdot D$) becomes redefined in terms of differential behavior via a simple explicit differential equation.

Differential Equation of Work

$$dW = F(x) \cdot dx$$

Finding the total work is simply a matter of continuous summing of the differential work elements dW from $x = a$ to $x = b$. The result is given by the definite integral

$$W = \int_a^b F(x)dx.$$

Figure 8.2 depicts the revised scenario with a non-constant force $F = F(x)$.

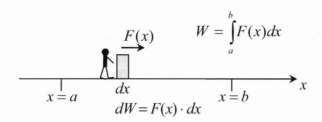

Figure 8.2: Work with Non-Constant Force

Ex 8.1: Find the total work performed by $F(x) = x^2 + 4$ as it is applied through the interval $[2,5]$. Let the x units be feet and, the F units, pounds force.

$$W = \int_a^b F(x)dx = \int_2^5 (x^2 + 4)dx =$$

$$\left[\frac{x^3}{3} + 4x\right]\Big|_2^5 = \left[\frac{5^3}{3} + 4 \cdot 5\right] - \left[\frac{2^3}{3} + 4 \cdot 2\right] =$$

$$\frac{117}{3} + 12 = 51 ft \cdot lb_f$$

Grabbing a work example from materials science, Hook's Law states that the force required to stretch or compress a string x units beyond its natural or resting length is given by $F(x) = kx$ where k is called the spring constant (a constant of proportionally). Hook's Law is only good for stretching lengths within what is called the *elastic limit*. Beyond the elastic limit, permanent deformation or set will take place; and Hook's law is no longer an appropriate mathematical model. Hook's Law also can be applied in other non-spring scenarios along as the material being studied is behaving in an elastic manner.

Ex 8.2: Suppose $10 in \cdot lb_f$ of work is required to stretch a spring from its resting length of 3 inches to a length of 5 inches. How much work is done in stretching the spring from 3 inches to a length of 9 inches? Assume that the stretching is such that Hook's Law applies. **Figure 8.3** diagrams the situation where the spring (drawn as a double line) is shown at its resting length of 3 inches.

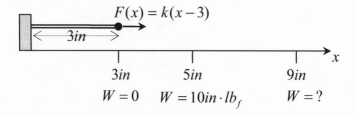

Figure 8.3: Hook's Law Applied to a Simple Spring

$\overset{1}{\mapsto}$: Determine spring constant k

$$W = \int_{3}^{5} k(x-3)dx = 10 \Rightarrow$$

$$\left[\frac{k(x-3)^2}{2} \right] \Big|_{3}^{5} = 10 \Rightarrow$$

$$k \left[\frac{(2)^2}{2} - 0 \right] = k[2] = 10 \Rightarrow$$

$$k = 5$$

****

$\overset{2}{\mapsto}$: Determine work needed to stretch from $3in$ to $9in$

$$W = \int_{3}^{9} 5(x-3)dx =$$

$$\left[\frac{5(x-3)^2}{2} \right] \Big|_{3}^{9} = 5 \left[\frac{6^2}{2} - 0 \right] =$$

$$5[18] =$$

$$90in \cdot lb_f$$

Now, let's really demonstrate the power of the differential equation as it is used to formulate an alternate expression for work in terms of kinetic energy. The Kinetic Energy (KE) for an object of mass m traveling at a constant velocity v is given by

---

Kinetic Energy

$$KE = \tfrac{1}{2} mv^2$$

Suppose an object of constant mass travels from $x = a$ to $x = b$ and, in doing so, increases its velocity as shown in **Figure 8.4**.

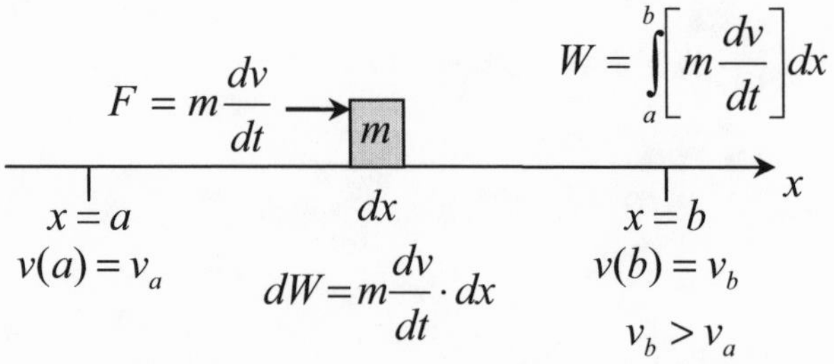

$$F = m\frac{dv}{dt} \longrightarrow \boxed{m} \qquad W = \int\limits_{a}^{b}\left[m\frac{dv}{dt} \right]dx$$

$$\xrightarrow{} x$$

$$\underset{v(a) = v_a}{x = a} \qquad dx \qquad \underset{\substack{v(b) = v_b \\ v_b > v_a}}{x = b}$$

$$dW = m\frac{dv}{dt}\cdot dx$$

Figure 8.4: Work and Kinetic Energy

Since velocity has increased from v_a to v_b, acceleration has occurred on the interval $[a, b]$. According to Newton's Second Law, this can not happen unless there has been an applied force. In the case of a constant mass m, this force is given by

$$F = (mv)' = m\frac{dv}{dt} = ma.$$

Substituting $m\dfrac{dv}{dt}$ into the differential equation for work

gives $dW = m\dfrac{dv}{dt}dx$. From the differential expression, the total work performed on the interval can be immediately obtained via

the definite integral $W = \int\limits_{a}^{b}m\dfrac{dv}{dt}dx$. On first glance, evaluating

this integral, which contains three differentials, seems to be an impossible task. Enter the power of a little differential rearrangement, which is a totally legitimate operation since differentials are algebraic quantities like any other algebraic

quantity. The first move is just noticing that velocity $v = \dfrac{dx}{dt}$, which

leads to

$$\frac{dv}{dt}dx = \frac{dx}{dt}dv = vdv$$

Then, in order to prepare for a definite integration with respect to x, the stated independent variable, we continue our transformation as follows

$$vdv = v\frac{dv}{dx}dx = vv'dx,$$

dividing and/or multiplying by appropriate differentials at will as long as we retain the algebraic balance. Tracing the whole development which equates work to a change in Kinetic Energy, we have:

$$W = \int_a^b m\frac{dv}{dt}dx \Rightarrow$$

$$W = \int_a^b m\frac{dx}{dt}dv = \int_a^b mvdv \Rightarrow$$

$$W = \int_a^b m[v]^1\frac{dv}{dx}dx = \int_a^b m[v]^1 v'dx \Rightarrow$$

$$W = \left[\frac{m\{v(x)\}^2}{2}\right]\Big|_a^b \Rightarrow$$

$$W = \tfrac{1}{2}mV_b^{\,2} - \tfrac{1}{2}mV_a^{\,2} = \Delta KE$$

Taking the analysis one step farther, suppose the object in **Figure 8.4** falls from a height h_a at $x = a$ to a height h_b at $x = b$ where $h_a > h_b$. *Note: Imagine an upward tilt at the left end.*

The only force acting is that due to gravity given by $F = mg$, which acts through a net distance of $h_a - h_b$. Hence, the work done on the object due to gravity is $W = mg[h_a - h_b]$ and must be equivalent to the change in Kinetic Energy experienced by the object. Thus,

$$mg[h_a - h_b] = \tfrac{1}{2}mv_b^{\,2} - \tfrac{1}{2}mv_a^{\,2} \Rightarrow$$
$$mgh_a + \tfrac{1}{2}mv_a^{\,2} = mgh_b + \tfrac{1}{2}mv_b^{\,2}$$

The term mgh is called potential energy (e.g. energy due to an elevated position), and the last equation expresses an energy conservation principle between two points a and b. It essentially states that the sum of kinetic and potential energy between any two points a and b on the path of the object remains constant no matter where a and b are located on the path of motion. The conservation principle holds as long as altitude changes are small with respect to the radius of the earth.

We shall formally state this conservation principle as expressed below, giving it Newton's name in honor of his Second Law of Motion, by which it was formulated through the power of calculus.

Newtonian Conservation of Energy Principle

In the absence of all external forces except that force due to gravity, the sum of the potential and kinetic energy for an object of constant mass m remains unchanged throughout the object's path of travel.

If a and b are any two points on the path of the object, and if the altitude changes are small when compared to the radius of the earth, this principle can be algebraically expressed (after canceling the m) as

$$gh_a + \tfrac{1}{2}v_a{}^2 = gh_b + \tfrac{1}{2}v_b{}^2 .$$

To summarize, the elegant expression of this magnificent energy-conservation principle on a macro scale was made possible by the careful algebraic manipulation of the tiny differential as it was found in Newton's Second Law. Granted, in our modern atomic age, Einstein's mass-to-energy conversion expression $E = mc^2$ can annihilate Newton's Conservation of Energy Principle in a flash. Nonetheless, his principle still reigns supreme, 300 years after its inception, as the right tool for most applications in our everyday and *earthbound* world.

Note: I am a native Ohioan. Two famous Ohioans whose historic altitude changes temporarily revoked the Newtonian Conservation of Energy Principle are John Glenn and Neil Armstrong. Ohio is also home to King's Island and Cedar Point, two famous amusement parks. And, for what thrill are these two amusement parks legendary?—roller coasters! Let's take a ride on an earthbound Rocket (or Beast) in our next example applying Newton's Conservation of Energy Principle.

Ex 8.3: A roller coaster descends through a vertical drop of 250 feet. At the apex, just before the drop, the velocity is $4.4\frac{ft}{s}$ ($3mph$). What is the coaster's velocity at the bottom of the drop ignoring rail and air friction?

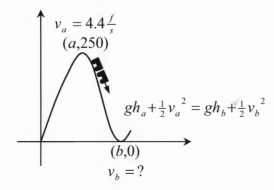

Figure 8.5: Newton Tames the Beast

Figure 8.5 shows our roller coaster ride and names the law that governs it. At the apex, let $h_a = 250\,ft$, $v_a = 4.4\frac{ft}{s}$. Since potential energy is a linear function of altitude, set $h_b = 0$ in order to get the needed drop or change. The appropriate units for the gravitational acceleration constant g in this example are $g = 32.2\frac{ft}{s^2}$. With these preliminaries in place, we can now calculate v_b via the formula $gh_a + \frac{1}{2}v_a^2 = gh_b + \frac{1}{2}v_b^2$.

$$gh_a + \tfrac{1}{2}v_a{}^2 = gh_b + \tfrac{1}{2}v_b{}^2 \Rightarrow$$

$$v_b = \sqrt{2gh_a + v_a{}^2} \Rightarrow$$

$$v_b = \sqrt{2gh_a + v_a{}^2} \Rightarrow$$

$$v_b = \sqrt{2(32.2)250 + (4.4)^2} \Rightarrow$$

$$v_b = \sqrt{16,100.10 + 19.36} \Rightarrow$$

$$v_b = \sqrt{16,119.36} = 126.96\tfrac{ft}{s} = 86.5mph$$

The final velocity of $86.5mph$ seems rather fast and scary. You might want to check the dimensional correctness of the above equality stream by ensuring both sides indeed reduce to the units of velocity (feet-per-second) throughout. Note that the conversion from feet-per-second to miles-per-hour is $88\tfrac{f}{s} = 60mph$.

Enough of roller coasters! I think it is time to escape the earth and go to the moon. Newton again shall be our guide via his Law of Universal Gravitation (remember the apple?). This law states that if two bodies of masses m_1 and m_2 are such that their respective centers of mass are r units apart, then the force due to gravity between them is given by

---

Newton's Law of Universal Gravitation

$$F = \frac{km_1 m_2}{r^2}$$

---

where k is called the *universal gravitational constant of proportionality*, given by $k = 6.67 x 10^{-11}\ \tfrac{N-m^2}{kg^2}$ in metric units.

To start our escape, suppose a rocket ship of mass m is poised for takeoff on planet earth, one such as the Saturn V that took the crew of Apollo 11 to the moon in July of 1969.

While sitting on planet earth, the rocket ship is experiencing a force due to gravity given by $F = -mg$. It is also positioned a distance R from the earth's center, which doubles as the earth's center of gravity. Let M be the mass of the earth. Then, after equating Newton's Law of Universal Gravitation to $F = -mg$, we have

$$\frac{kmM}{R^2} = -mg \Rightarrow k = -\frac{R^2 g}{M}.$$

Substituting this expression for k back into Newton's Law of Universal Gravitation gives after some quick algebraic rearranging

$$F = -mg\left(\frac{R}{r}\right)^2.$$

Let's pause here for a second and examine the above. Notice that the force due to the earth's gravity upon an object of mass m depends on how far that object is away from the earth's center. When $r = R$, corresponding to the earth's surface where $R \cong 4000$ miles, the force reduces to the good 'ol familiar $F = -mg$. For distances h above the earth's surface, the force can be written as

$$F = -mg\left(\frac{R}{R+h}\right)^2 = -mg\left(\frac{1}{1 + \{\frac{h}{R}\}}\right)^2.$$

Hence, when h is small compared to R, whether be it the altitudinal extent of a roller coaster ride or a supersonic trip in the Concorde, the force F is approximately equal to $-mg$. In cases such as these, the Newtonian Conservation of Energy Principle (as stated by $gh_a + \frac{1}{2}v_a^2 = gh_b + \frac{1}{2}v_b^2$) applies quite well. But, we are going to the moon as shown in **Figure 8.6**, and our h values will be large. With this in mind, how does the Newtonian Conservation of Energy Principle change?

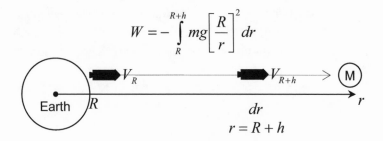

Figure 8.6: From Earth to the Moon

The answer is that Newton's Energy Conservation Principle really doesn't change at all if we recall that it was developed from the more basic relationship

$$W = \tfrac{1}{2}mV_b^{\,2} - \tfrac{1}{2}mV_a^{\,2} = \Delta KE \;.$$

What has changed is the nature of the applied gravity force, which now varies as a function of r. The change in kinetic energy for our rocket still equals the work done *against the force of gravity*. After a high initial velocity V_R, achieved after a hundred or so miles in altitude, one would expect that the velocity would decay with distance as the rocket seeks to free itself from the drag force of gravity. The change in kinetic energy is equal to the work done against this force, which is now a function of the distance r from the center of the earth. Using a definite integral to express the total work done on the interval $[R, R+h]$ in moving an object of mass m against the force of gravity, we have

$$W = -\int_{R}^{R+h} mg\left[\frac{R}{r}\right]^2 dr \;.$$

Equating work to the corresponding change in kinetic energy gives

$$\tfrac{1}{2}mV_{R+h}^{\,2} - \tfrac{1}{2}mV_R^{\,2} = -\int_{R}^{R+h} mg\left[\frac{R}{r}\right]^2 dr \;.$$

After canceling the m and evaluating the definite integral, the last expression reduces to

$$\tfrac{1}{2}V_{R+h}^{2} - \tfrac{1}{2}V_{R}^{2} = g\left[\frac{R^2}{R+h} - \frac{R^2}{R}\right] \Rightarrow$$

$$\tfrac{1}{2}V_{R+h}^{2} - \tfrac{1}{2}V_{R}^{2} = -g\frac{Rh}{R+h} = -g\frac{R}{1+\frac{R}{h}}.$$

Continuing, what is the initial velocity V_R needed at $r = R$ to guarantee a forward velocity V_{R+h} at $r = R+h$? Solving the last equation for V_R gives the answer:

$$V_R = \sqrt{\left[V_{R+h}^{2} + \frac{2gR}{1+\frac{R}{h}}\right]}$$

Now, escape velocity is defined as that initial velocity needed to guarantee some forward velocity $V_{R+h} > 0$ as $h \to \infty$. The last condition can be expressed as $\lim_{h\to\infty}[V_{R+h}] = 0$, the first limit encountered since Chapter 5. Applying the limit gives

---

The Equation for Escape Velocity

$$V_{Escape} = \lim_{h\to\infty}\left\{\sqrt{\left[V_{R+h}^{2} + \frac{2gR}{1+\frac{R}{h}}\right]}\right\} \Rightarrow$$

$$V_{Escape} = \sqrt{2gR}.$$

In the case of planet earth, we have $g = 32.2\frac{ft}{s^2}, R = 21{,}120{,}000\,ft$.

Thus $V_{Escape} = \sqrt{2gR} = 36{,}937.17\frac{ft}{s} \cong 7mps$ (miles per second).

Ex 8.4: What initial velocity V_R is needed to insure that an interplanetary space probe has a forward velocity of $1.0mps$ at the point of the moon's orbit?

We shall use $R = 21{,}120{,}000\,ft$ and $h = 240{,}000\,miles$, which equals $1{,}267{,}200{,}000\,ft$. Hence the $\dfrac{R}{h}$ ratio is $\dfrac{R}{h} = .016667$.

Letting $V_{R+h} = 5280\frac{ft}{s}$ and $g = 32.2\frac{ft}{s^2}$, we obtain

$$V_R = \sqrt{\left[(5280)^2 + \frac{2(32.2)(21{,}120{,}000.0)}{1+.016667} \right]} \Rightarrow$$

$$V_R = 36{,}955\,\tfrac{ft}{s} = 6.999mps$$

Notice that this last answer is only slightly larger than the value for V_{Escape} given on the previous page. This means that a space probe propelled to V_{Escape} in the very early stages of flight will still have a forward velocity of $1.0mps$ at the point of the moon's orbit.

Ex 8.5: Find V_{Escape} for the moon.

For the moon, $R = 1088\,miles$ and $g = 5.474\frac{ft}{s^2}$. Thus, we have

$$V_{Escape} = \sqrt{2gR} = 7930\tfrac{ft}{s} = 1.502mps.$$

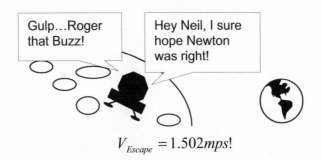

$$V_{Escape} = 1.502mps!$$

Figure 8.7: Just Before Lunar Takeoff

8.2.2) Jacob Bernoulli's Multi-Purpose Differential Equation

Jacob Bernoulli (1654-1705) was nestled in between the lifetimes of Leibniz and Newton. Being about 10 years younger than either of these two independent co-developers of Calculus, Jacob was, the first of many to continue the tradition of 'standing on the shoulders of giants'. One of Jacob's greatest contributions to mathematics *and physics* was made in the year 1696 when he found a solution to the differential equation below, which now bears his name.

The Bernoulli Differential Equation

$$\frac{dy}{dx} = f(x)y + g(x)y^n,$$

Bernoulli's equation is neither explicit nor immediately separable into the form $P(y)dy = Q(x)dx$. Hence, after a flash of pure genius, Jacob rewrote the equation as

$$y^{-n}y' = f(x)y^{1-n} + g(x).$$

He then made the following *change-of-variable* substitution

$$z = y^{1-n} \Rightarrow z' = (1-n)y^{-n}y'$$

to obtain

$$\frac{z'}{1-n} = f(x)z + g(x).$$

In the next subsection, we will solve a specific example of the above differential equation using Bernoulli's change-of-variable technique for the case $n = 2$. In this subsection, we will solve the general Bernoulli equation for the case $n = 0$, which reduces to (after replacing z with y)

$$\frac{dy}{dx} = f(x)y + g(x).$$

Proceeding stepwise with Bernoulli's general solution:

$\overset{1}{\mapsto}$: Let $F(x)$ be such that $F'(x) = -f(x)$.

$\overset{2}{\mapsto}$: Form the function $e^{F(x)}$ (called an integrating factor)

$\overset{3}{\mapsto}$: Multiply both sides of $\dfrac{dy}{dx} = f(x)y + g(x)$ by $e^{F(x)}$

$$\Rightarrow e^{F(x)}\left[\frac{dy}{dx}\right] = e^{F(x)}[f(x)y + g(x)] \Rightarrow$$

$$e^{F(x)}\left[\frac{dy}{dx}\right] + e^{F(x)}[-f(x)]y = e^{F(x)} \cdot g(x)$$

Notice that the left-hand side of the last equality is the result of differentiating the product $e^{F(x)} \cdot y$:

$$\frac{d}{dx}\left[e^{F(x)} \cdot y\right] = e^{F(x)}\left[\frac{dy}{dx}\right] + e^{F(x)}[-f(x)]y = e^{F(x)} \cdot g(x).$$

Consequently, the term $e^{F(x)}$ *is known as an integrating factor because it allows the integration shown in Step 4 to take place.*

$$\overset{4}{\mapsto} : \frac{d}{dx}\left[e^{F(x)} \cdot y\right] = e^{F(x)} \cdot g(x) \Rightarrow$$

$$e^{F(x)} \cdot y = \int e^{F(x)} \cdot g(x)dx + C \Rightarrow$$

$$y = e^{-F(x)}\left[\int e^{F(x)} \cdot g(x)dx\right] + Ce^{-F(x)} \therefore$$

Admittedly, the final equality is a rather atrocious looking expression, but Bernoulli's 300-year-old masterpiece will always give us the right solution if we faithfully follow the solution process embodied in the formula.

Ex 8.6: Solve the differential equation $\dfrac{dy}{dt} = ay + b : y(0) = y_0$

where a, b are constants, and t is the independent variable.

The given differential equation is Bernoulli in form. In practice, we solve using the following process (which is slightly modified from the process shown for the general solution)

$\overset{1}{\mapsto}: \dfrac{dy}{dt} = ay + b \Rightarrow \dfrac{dy}{dt} - ay = b$

$\overset{2}{\mapsto}:$ Create $e^{\int -a\,dt} = e^{-at}$, the integrating factor

$\overset{3}{\mapsto}: e^{-at} \cdot \dfrac{dy}{dt} - ae^{-at} \cdot y = e^{-at} \cdot b \Rightarrow$

$\dfrac{d}{dt}\left[e^{-at} \cdot y\right] = e^{-at} \cdot b \Rightarrow$

$e^{-at} \cdot y = \int e^{-at} \cdot b\,dt + C \Rightarrow$

$e^{-at} \cdot y = \tfrac{-b}{a} \int e^{-at} \cdot (-a) \cdot dt + C \Rightarrow$

$e^{-at} \cdot y = \tfrac{-b}{a} e^{-at} + C \Rightarrow$

$y = \tfrac{-b}{a} + Ce^{at}$

$\overset{4}{\mapsto}:$ Apply the initial condition $y(0) = y_0$.

$y_o = \tfrac{-b}{a} + Ce^{a \cdot 0} \Rightarrow C = y_0 + \tfrac{b}{a}$

$\overset{5}{\mapsto}:$ Substitute the value for C and finalize the solution.

$y = \tfrac{-b}{a} + \left[y_0 + \tfrac{b}{a}\right]e^{at} \Rightarrow$
$y = y_0 e^{at} + \tfrac{b}{a}\left[e^{at} - 1\right] \therefore$

Note: The final result is going to prove itself extremely useful throughout the remainder of this section and the next section.

The reason that Bernoulli's Differential Equation is so important is that it "pops up" (albeit with various values of the exponent n) in a variety of diverse situations where physics is being applied from free fall with atmospheric drag to elementary electric circuit theory.

For example, if we assume atmospheric drag is a primary player affecting the motion of a falling body of mass m, then the governing equation is

$$-m\frac{dv}{dt} = -mg + kv^n.$$

The term kv^n is the atmospheric drag force that acts in opposition to the falling motion. For some objects, this drag term is proportional to the square of the velocity. For others, it is proportional to the first power; and yet for others, whatever empirical testing supports. When $n = 1$, the equation is Bernoulli in form and solvable by Bernoulli's method. The example below is a free-fall problem where $n = 1$.

Ex 8.7: A 160 pound sky diver bails out of an airplane cruising at 7000 feet. The free-fall phase of the jump lasts 15 seconds. Find the velocity and position of the sky diver at the end of free fall.

The governing differential equation for a human body in free fall is given by $-m\frac{dv}{dt} = -mg + (.5) \cdot v^1$ where the drag term $(.5) \cdot v^1$ has been empirically deduced via years of data. Continuing:

$$-m\frac{dv}{dt} = -mg + (.5) \cdot v^1 \Rightarrow$$

$$\frac{dv}{dt} = -\frac{v}{2m} + g : v(0) = 0\tfrac{ft}{s}, y(0) = 7000\,ft$$

Since the above differential equation is Bernoulli in form, one can immediately write the solution using the result from **Ex 8.6** with $a = \frac{-1}{2m} = -0.100625, b = 32.2, v(0) = 0$. The weight of the sky diver must first be converted from pounds force lbf to pounds mass lbm, which is done by dividing by g. Continuing:

$$v(t) = 320\left[e^{-0.100625t} - 1\right] \Rightarrow$$

$$v(20) = 320\left[e^{-0.100625(15)} - 1\right] = -249\tfrac{ft}{s} = -170mph$$

where the minus sign indicates the direction of fall is towards the earth.

From the previous expression, notice that velocity will never exceed $-320\frac{ft}{s} = -218mph$. The velocity $-320\frac{ft}{s}$ is known as the terminal velocity and is the velocity at which gravitational and atmospheric drag forces balance. Solving for $y(t) = \int v(t)dt$, we have:

$$\overset{1}{\mapsto} : v(t) = \frac{dy}{dt} = 320\left[e^{-0.100625t} - 1\right] \Rightarrow$$

$$dy = 320\left[e^{-0.100625t} - 1\right]dt \Rightarrow$$

$$y = y(t) = \int 320\left[e^{-0.100625t} - 1\right]dt + C \Rightarrow$$

$$y(t) = -3180e^{-0.100625t} - 320t + C$$

$$\overset{2}{\mapsto} : y(0) = 7000 \Rightarrow C = 10180$$

$$\overset{3}{\mapsto} : y(t) = 10180 - 3180e^{-0.100625t} - 320t$$

$$\overset{4}{\mapsto} : y(15) = 10180 - 3180e^{-0.100625(15)} - 320(15) \Rightarrow$$

$$y(15) = 4677 ft$$

Note: a very good model for the parachute-open portion of the sky dive is $-m\frac{dv}{dt} = -mg + (.42) \cdot v^2$, *which is not Bernoulli in form. However, it is still solvable via the slightly more advanced methods presented in an introductory course on differential equations.*

In general, suppose the *drag coefficient* is k for the term kv^1 and the mass of the object is m. Then $v(t) = \frac{-gm}{k}[e^{\frac{-kt}{m}} - 1]$ and the terminal velocity is given by $\frac{-gm}{k}$. Hence, for a $160\,lbf$ sky driver, a drag coefficient of $k \geq 8$ would be needed in order to keep the terminal velocity below $-20\frac{ft}{s}$.

Our next example, one also requiring the solution of a Bernoulli equation, is taken from elementary circuit theory.

Ex 8.8: A simple electric circuit consists of a resistance R, an inductance L, and an electromotive force, E, connected in series as shown **Figure 8.8**. If the switch, S, is thrown at time $t = 0$, express the current i as a function of time.

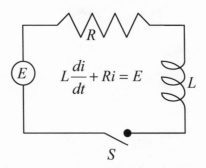

Figure 8.8: A Simple Electric Circuit

Once the switch is thrown, the governing differential equation is given by

$$L\frac{di}{dt} + Ri = E : i(0) = 0,$$

derivable using Kirchoff's laws for electric circuits. The above equation can be rewritten as

$$\frac{di}{dt} = -\frac{R}{L}i + \frac{E}{L} : i(0) = 0.$$

This should be immediately recognized as Bernoulli in form and matching the pattern of Ex 8.6 with $a = \frac{-R}{L}, b = \frac{E}{L}$. Hence the solution is given by

$$i(t) = \frac{E}{R}\left[1 - e^{-\frac{Rt}{L}}\right].$$

With time, the current i approaches a *steady state value* of $\frac{E}{R}$, which is akin to terminal velocity in the free-fall problem.

We end this subsection with an example coming from mass flow, an everyday earthbound problem, having a somewhat sophisticated and unique Bernoulli solution.

Ex 8.9: A $1000gal$ water tank is holding $100gal$ of pure water. Brine containing $2lbm$ of salt per gallon starts to flow into the tank at a steady rate of $4\frac{gal}{min}$. Concurrently, the brine mixture starts to flow out of the bottom of the tank at a steady rate of $2\frac{gal}{min}$. Assuming thorough mixing throughout the salinization process, write an expression for s, the number of pounds of salt in the tank, as a function of time t.

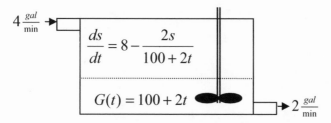

Figure 8.9: Dynamic Brine Tank

Figure 8.9 shows the brine tank in process. After t minutes of operation, the amount of gallons in the tank is given by the $G(t) = 100 + (4-2)t = 100 + 2t$, a simple conservation of volume expression for an incompressible liquid. Since the tank's capacity is 1000 gallons, we can immediately find the total time of operation by solving the equation

$$1000 = 100 + 2t \Rightarrow t = 450 \min(7.5hr).$$

During any one instant of time, *the time rate of change of salt in the tank is equal to the rate that salt is flowing into the tank minus the rate that the salt is exiting the tank.* The expression *time rate of change* clues us to the fact that we are dealing with differentials and hence, a differential equation. The inflow of salt is equal to $(2\frac{lbm}{gal}) \cdot (4\frac{gal}{min}) = 8\frac{gal}{min}$. The outflow expression is a bit more difficult to obtain, but workable.

285

Let s be the pounds of salt in the tank at time t. The number of gallons in the tank associated with this same time is $100 + 2t$.

Thus, the concentration of salt ($\frac{lbm}{gal}$) at time t is given by $\dfrac{s}{100 + 2t}$.

Finally, since the liquid is flowing out at a steady rate of $2\frac{gal}{min}$, the amount of salt being carried out with the exit flow is $\dfrac{2s}{100 + 2t}$.

Inserting the various mathematical expressions into the rate-equality statement (italicized) leads to the following Bernoulli-in-form differential equation

$$\frac{ds}{dt} = 8 - \frac{2s}{100 + 2t} : s(0) = 0.$$

Solving:

$\overset{1}{\mapsto} : \dfrac{ds}{dt} = 8 - \dfrac{2s}{100 + 2t} \Rightarrow$

$\dfrac{ds}{dt} + \dfrac{2s}{100 + 2t} = 8$

$\overset{2}{\mapsto} :$ Let $e^{\ln(100+2t)} = 100 + 2t$ be the integrating factor.

$\overset{3}{\mapsto} : (100 + 2t)\left[\dfrac{ds}{dt} + \dfrac{2s}{100 + 2t}\right] = 8(100 + 2t) \Rightarrow$

$(100 + 2t)\dfrac{ds}{dt} + 2s = 8(100 + 2t) \Rightarrow$

$\dfrac{d}{dt}[(100 + 2t)s] = 8(100 + 2t) \Rightarrow$

$(100 + 2t)s = 4\int(100 + 2t) \cdot 2 \cdot dt + C \Rightarrow$

$(100 + 2t)s = 2(100 + 2t)^2 + C \Rightarrow$

$s = s(t) = 2(100 + 2t) + \dfrac{C}{100 + 2t}$

$\overset{4}{\mapsto}$: Apply the initial condition $s(0) = 0$.

$$0 = 2(100 + 2 \cdot 0) + \frac{C}{100 + 2 \cdot 0} \Rightarrow$$

$$C = -20,000$$

The final solution for the amount of salt (lbm) in the tank at time t is given by

$$s(t) = 2(100 + 2t) - \frac{20,000}{100 + 2t} .$$

From the above, we can immediately calculate the amount of salt in the tank at the instant that the tank is filled.

$$s(450) = 2(100 + 2 \cdot 450) - \frac{20,000}{100 + 2 \cdot 450} = 1980 lbm$$

We can also determine the concentration of salt at that point $c = \frac{1980 lbm}{1000 gal} = 1.98 \frac{lbm}{gal}$, which almost equal to the concentration of the inflow. So in a real sense, we are near *steady state* with respect to operation of the salinization process.

Finally, suppose we wanted to stop the process when the salt concentration level reached $1 \frac{lbm}{gal}$. To find when this would be, set

$$c(t) = \frac{s(t)}{100 + 2t} = 2 - \frac{20,000}{(100 + 2t)^2} = 1 .$$

Solving for t, one obtains

$$-\frac{20,000}{(100 + 2t)^2} = -1 \Rightarrow (100 + 2t)^2 = 20,000 \Rightarrow t = 20.7 \min .$$

Thus, at $t = 20.7 \min$, there is $141.4 lbm$ of salt in the tank and an equivalent number of gallons.

8.2.3) Growth and Decay Laws

We end this section with a very short discussion of growth and decay laws. If the time-rate-of-change for a quantity y is proportional to the amount of y present, then one can easily formulate this verbal statement by the differential equation

$$\frac{dy}{dt} = ky.$$

The constant k is called the growth constant of proportionality. If $k > 0$, then the above is a *growth law*, and if $k < 0$, a *decay law*. The solution in either case is given by

---

Solution to Growth/Decay Differential Equation

If $\dfrac{dy}{dt} = ky : y(0) = y_0$, then $y = y_0 e^{kt}$.

Notice that the above differential equation is yet another example—a very simple one—of a Bernoulli equation.

Though simple, the differential equation $\dfrac{dy}{dt} = ky$ can be used to describe a variety of phenomena such as: growth of money, inflation (decay of money), planned decay of money (i.e. the paying off of an installment loan over time), half-life of radioactive substances, and population growth of living organisms. In the remainder of this section, we shall now look at half life and population growth, 'saving our money' for the final section in this, the last major chapter of the book.

Ex 8.10: Radium decays exponentially according to $y = y_0 e^{kt}$ and has a half-life of 1600 years. This means that if y_0 is the initial amount of radioactive substance, then $0.5 y_0$ remains at $t = 1600$. Armed with this information, find the time t when the amount of remaining radioactive substance is $0.1 y_0$.

$\mapsto^{1} : (0.5)y_0 = y_0 \cdot e^{1600k} \Rightarrow$

$0.5 = e^{1600k} \Rightarrow$

$k = \frac{-\ln 2}{1600} \Rightarrow$

$y(t) = y_0 e^{(\frac{-\ln 2}{1600})t} = y_0 (2)^{(\frac{-t}{1600})}$

$\mapsto^{2} : (0.1)y_0 = y_0 (2)^{(\frac{-t}{1600})} \Rightarrow$

$(2)^{(\frac{-t}{1600})} = (0.1) \Rightarrow$

$(2)^{(\frac{t}{1600})} = 10 \Rightarrow$

$\frac{t}{1600} Log(2) = Log(10) = 1 \Rightarrow$

$t = 5315 \, years$

In the real world, biological populations do not experience unbounded growth as suggested by the simple exponential function $y = y_0 e^{kt} : k > 0$. Hence, the corresponding differential equation $\frac{dy}{dt} = ky$ needs to be modified in order to reflect the fact that the environment hosting the population has a finite carrying capacity $L > y_0 > 0$. This is easily done by turning the growth constant k into a variable that collapses to zero as the independent variable approaches the carrying capacity L. The easiest way to express this relationship is by

---

The Law of Logistic Growth

$$\frac{dy}{dt} = k(L - y)y : y(0) = y_0 : k > 0 : L > y_0 > 0$$

Ex 8.11: Solve the differential equation governing the Law of Logistic Growth and plot the general solution on an appropriate $x - y$ coordinate system.

First, we will rewrite the differential equation governing the Law of Logistic Growth as follows

$$\frac{dy}{dt} = kLy - ky^2 \, ,$$

which is Bernoulli-in-form for the case $n = 2$.

$\overset{1}{\mapsto}$: Divide both sides by y^2

$$\left[\frac{1}{y^2}\right]\frac{dy}{dt} = kL\left[\frac{1}{y}\right] - k \Rightarrow$$

$$\left[\frac{1}{y^2}\right]\frac{dy}{dt} - kL\left[\frac{1}{y}\right] = -k$$

$\overset{2}{\mapsto}$: Let $z(t) = \dfrac{1}{y(t)}$, which implies that $\dfrac{dz}{dt} = \left[\dfrac{-1}{y^2}\right] \cdot \dfrac{dy}{dt}$.

$\overset{3}{\mapsto}$: $\left[\dfrac{1}{y^2}\right]\dfrac{dy}{dt} - kL\left[\dfrac{1}{y}\right] = -k \Rightarrow$

$$-\frac{dz}{dt} - kLz = -k \Rightarrow \frac{dz}{dt} + kLz = k$$

$\overset{4}{\mapsto}$: Form the integrating factor e^{kLt}

$\overset{5}{\mapsto}$: $\dfrac{dz}{dt} + kLz = k \Rightarrow$

$$e^{kLt}\left[\frac{dz}{dt} + kLz\right] = ke^{kLt} \Rightarrow$$

$$\frac{d}{dt}\left[e^{kLt}z\right] = ke^{kLt} \Rightarrow$$

$$e^{kLt}z = \int ke^{kLt}\,dt + C \Rightarrow$$

$$z = z(t) = \frac{1}{L} + Ce^{-kLt}$$

$\mapsto$: Replace $z(t)$ with $\dfrac{1}{y(t)}$ and solve for $y(t)$.

$$z(t) = \frac{1}{y(t)} = \frac{1}{L} + Ce^{-kLt} \Rightarrow$$

$$y(t) = \frac{L}{1 + CLe^{-kLt}}$$

$\mapsto$: Apply $y(0) = y_0$ in order to solve for C.

$$z(t) = \frac{1}{y(t)} = \frac{1}{L} + Ce^{-kLt} \Rightarrow$$

$$y(0) = y^0 = \frac{L}{1 + CLe^{-kL(0)}} \Rightarrow$$

$$y_0 = \frac{L}{1 + CL} \Rightarrow$$

$$C = \frac{L - y_0}{Ly_0}$$

$\mapsto$: Substitute for C and simplify to complete the solution.

$$y(t) = \frac{Ly_0}{y_0 + (L - y_0)e^{-kLt}} \quad \therefore$$

To summarize, we have just completed the most intensive (in terms of the number of steps) solution process in this book by solving a Bernoulli differential equation for the case $n = 2$ utilizing Bernoulli's change-of-variable technique. The particular equation we solved was very practical in the sense that it modeled the Law of Logistic Growth. This law governs the overall growth phenomena for biological populations residing in an environment having an underlying carrying capacity L.

Moving on to the finishing touch, **Figure 8.10** is a graph of $y(t)$ where $L > y_0 > 0$ and $t \geq 0$.

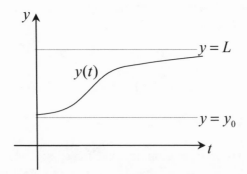

Figure 8.10: Graph of Logistic Growth Equation

Figure 8.10 shows $y(t)$ exhibiting exponential-like behavior early in the growth process. This behavior rapidly slows down and the curve starts bending back once the ceiling imposed by the carrying capacity *kicks in*—so to speak.

$$\int_{a}^{b} \overset{\cdot\cdot}{\cup} dx$$

Section Exercises

1. The sun has a mean radius of $431{,}852$ miles and $V_{Escape} = 2{,}027{,}624.05\frac{ft}{s}$ at the sun's surface. Calculate the escape velocity V_{Escape} needed to break free of the solar system where the starting point is planet Earth. Earth's mean distance from the sun is $93{,}000{,}000$ miles.

2. A 240 grain slug ($1 grain = 60 mg$) impacts a ballistic jell at a velocity $v = 529\frac{m}{s}$ and comes to rest at $t = 0.01s$. The governing equation of the motion is $m\frac{dv}{dt} = -k\sqrt{v}$. Find the value of the constant k, the force of impact, and the penetration depth.

3. Solve the following Bernoulli-in-form differential equations

a) $\dfrac{dy}{dx} = 2y + 1 : y(0) = 1$ b) $\dfrac{dy}{dx} = 2y + y^2 : y(0) = 1$

c) $\dfrac{dy}{dx} = 2y + y^3 : y(0) = 1$ d) $\dfrac{dp}{dt} = rp + c_0 e^{\alpha t} : p(0) = p_0$

4. Fifty deer are released in an isolated state park having a total carrying capacity of 800 individuals. Five years later the deer population is 300 individuals. Assuming that the Law of Logistic Growth applies, find the time needed to achieve a deer population that is 75% of the stated carrying capacity.

5. The work required to stretch a spring from a resting length of $5in$ to a length of $10in$ is $120in \cdot lbf$. How much work is required to stretch the spring from $10in$ to $15in$? Assume Hook's Law applies.

6. If a radioactive substance has a half-life of 10 years, find the time needed to reduce original radioactivity level by 95%. How long should this substance remain buried in underground crypts if a radiological safe level is considered to be 0.1%.

7. Determine the vertical drop needed for a roller coaster to achieve a speed of $110mph$ at the low point of the curve. Assume that the roller coaster rounds the apex at a speed of $2mph$.

8. Let a retarding force be defined by $F(x) = x^2 - 10x$ on the interval $[0,10]$. Find the total work done against the force in moving from $x = 1$ to $x = 10$. Find the point in the interval where the rate-of-change of applied force with respect to change in forward distance is a maximum.

8.3) Applications in Finance

In this, our final section, we are going to explore three typical financial concerns common to the modern American family: inflation, investments, and installments (as in payments). Surprisingly, we shall find that Jacob Bernoulli holds the key to easy formulation of the key parameters governing all three economic phenomena.

8.3.1) Inflation

Inflation is decay of money or, alternately, the purchasing power provided by money. Inflation is characterized by a yearly inflation rate i of so-many-percent per year and is assumed to act continuously throughout the year as prices go up weekly, monthly, etc. *If $P(t)$ is our present purchasing power, then the time-rate-of-change of this purchasing power $\dfrac{dP}{dt}$ is proportional to the amount of purchasing power currently present.* The constant of proportionality is the inflation rate i, which, when actually applied, is negative since inflation nibbles away at current purchasing power when current purchasing power is moved forward in time. Translating the italicized statement, continuous inflation is easily modeled by the following Bernoulli-in-form differential equation.

Differential Equation for Continuous Inflation

$$\frac{dP}{dt} = -iP : P(0) = P_0 \Rightarrow$$

$$P = P_0 e^{-it}$$

The constant i is the inflation rate.

When projecting the purchasing power P forward in time, an economist will call P the future value FV. The starting or current value P_0 is then referred to as the present value PV.

With this change in nomenclature, the inflation equation becomes

> **Present-Value-to-Future-Value**
> **Continuous Inflation Equation**
>
> $$FV = PVe^{-it}$$

Ex 8.12: Find the future value of $100.00 ten years from now if the annual inflation rate is $2.5\frac{\%}{year}$.

As written the inflation number is clearly seen as a proportionality constant for rate equivalency. Rates are all about differential change ratios and rate equivalencies are all about the corresponding differential equations.

$$FV = \$100.00e^{-0.025(10)} = \$77.80$$

So, rephrasing, $100.00 today will be worth (or have a buying power of) $77.80 ten years from now. The underlying assumption is an annual inflation rate equal to a steady 2.5% throughout the ten-year time period. As most baby-boomers well know, this inflation rate can be significantly higher at times, in which case the $77.80 figure will need to be readjusted downward.

Note: Calculus-based economic applications exploded after 1900, over 200 years after calculus was first invented.

Ex 8.13: Investment counselors say that if one wants to retire today, they need to have at least $1,000,000.00 in retirement savings. Translate this advice into future terms for our sons and daughters who will be retiring 40 years from now. Assume an average annual inflation rate of $3\frac{\%}{year}$.

In short, one wants to have the same buying power 40 years from now as $1,000,000.00 provides today. Or, looking at it in reverse, what is the present value associated with $1,000,000.00 40 years from now?

The last statement leads to

$$FV = \$1,000,000.00 = PVe^{-0.03(40)} \Rightarrow$$
$$PV = \$1,000,000.00e^{0.03(40)} = \$3,320,000.00$$

Thus, our son and daughters will need to plan on having about 3.3 million dollars in investments at the time of their retirement.

Next, we shall look at staged inflation problem where the inflation rate actually changes during the time-period examined. The technique utilized to accommodate this inflation change is easily adapted to all three types of finance problems in this section.

Ex 8.13: Find the future value of $\$100.00$ ten years from now if the annual inflation rate is $2.5\frac{\%}{year}$ for the first seven years and $4.1\frac{\%}{year}$ for the last three years.

$\overset{1}{\mapsto}$: Calculate the decline in value for the first seven years.
$$FV = \$100.00e^{-0.025(7)} = \$83.95$$

$\overset{2}{\mapsto}$: Use the $\$83.95$ as input into the second three-year stage.
$$FV = \$83.95e^{-0.041(3)} = \$74.23$$

8.3.2) Investments

Principle grows by two mechanisms: 1) the application of an interest rate to the current principle and 2) additional deposits. Principle growth can be marvelously summarized via the following italicized statement: *the time-rate-of-change of principle is proportional to the principle currently present plus the rate at which additional principle is added,* which can be mathematically rendered as

$$\frac{dP}{dt} = r(t)P + c(t) : P(0) = P_0 .$$

The above differential equation is Bernoulli-in-form with a non-constant interest rate $r(t)$ and principle addition rate $c(t)$.

In principle-growth problems, time is traditionally measured in years. Hence rates are expressed as so much per year. To start our exploration of principle growth, we will assume a constant annual interest rate r and a constant annual principle addition rate c. Under these restrictions, the differential equation governing continuous principle growth is easily solvable (**Ex 8.6**). We have

---

Differential Equation for Continuous Principle Growth

$$\frac{dP}{dt} = rP + c_0 : P(0) = P_0 \Rightarrow$$

$$P(t) = P_0 e^{rt} + \frac{c_0}{r}(e^{rt} - 1)$$

The parameters r, c_0 are assumed to be constant.

One might say, *what about compounding? New interest is only added to my savings account quarterly or semi-annually and not continuously as suggested by the above model.* To answer, it turns out that the continuous interest model provides a very accurate estimate of a final balance whenever the number of compounding periods N is such that $N \geq 4$ in any one year. Also, don't forget that principle growth is achieved by two mechanisms: 1) periodic application of the annual interest rate (compounding) and 2) direct addition to the principle through an annual contribution. For most of us, this annual contribution is made through steady metered installments via payroll deduction. The installments may be weekly, biweekly or monthly (e.g. members of the U.S. Armed Forces). In all three cases, the *compounding action* due to direct principle addition far exceeds the minimum of 4 compounding periods per year.

Note: Refer to Appendix C for equivalent formulas (to those presented in this section) based on a discrete number of annual compounding periods. Additionally, you might want to compare results based on continuous compounding with more traditional results based on a discrete number of compounding periods.

Returning to $P(t) = P_0 e^{rt} + \dfrac{c_0}{r}(e^{rt} - 1)$, notice that the expression consists of two distinct terms. The term $P_0 e^{rt}$ corresponds to the principle accrued in an interest-bearing account given an initial lump-sum investment P_0 over a time period t at a constant interest rate r. We first saw this formula in Chapter 4 when we explored the origin of the number called e. We now see it again, naturally arising from a simple—but most elegant—differential equation. Likewise, the term $\dfrac{c_0}{r}(e^{rt} - 1)$ results from direct principle addition via annual metered contributions into the same interest-bearing account. If either of the constants P_0 or c_0 is zero, then the corresponding term drops away from the overall expression.

Our first example in this subsection is a two-stage investment problem.

Ex 8.14: You inherit $\$12,000.00$ at age 25 and immediately invest $\$10,000.00$ in a corporate-bond fund paying $6\frac{\%}{year}$. Five years later, you roll this account over into a solid stock fund (whose fifty-year average is $8\frac{\%}{year}$) and start contributing $\$3000.00$ annually. **A)** Assuming continuous and steady interest, how much is this investment worth at age 68? **B)** What percent of the final total was generated by the initial $\$10,000.00$? **C)** What is the present value PV of this final total at age 25 assuming an annual inflation rate of $3\frac{\%}{year}$?

A) $\overset{1}{\longmapsto}$: In the first five years, the only growth mechanism in play is that made possible by the initial investment of $\$10,000.00$:

$$P(5) = \$10,000.00 e^{0.06(5)} = \$13,498.58.$$

$\overset{2}{\longmapsto}$: The output from Stage 1 is now input to Stage 2 where both growth mechanisms are acting for an additional 38 years.

$$P(38) = 13,498.58e^{0.08(38)} + \frac{3000}{0.08}(e^{0.08(38)} - 1) \Rightarrow$$

$$P(38) = \$148,797.22 + \$375,869.11 \Rightarrow$$

$$P(38) = \$528,666.34$$

B) The % of the final total accrued by the initial $\$10,000.00$ is

$$\frac{\$148,792.22}{\$528,666.34} = .281 = 28.1\%$$

Note: The initial investment of $\$10,000.00$ *is generating* 28.1% *of the final value even though it represents only* 8% *of the overall investment of* $\$124,000.00$. *The earlier a large sum of money is inherited or received by an individual, the wiser it needs to be invested; and the more it counts later in life.*

C) And now, we have the bad news.

$$PV = \$528,666.34e^{-0.03(43)} = \$145,526.40$$, which is nowhere near the suggested figure of $\$1,000,000.00$.

Holding the annual contributions at a steady rate over a period of 38 years is not a reasonable thing to do. As income grows, the corresponding retirement contribution should also grow. A great model for this is

$$\frac{dP}{dt} = rP + c_0 e^{\alpha t} : P(0) = P_0,$$

where the annual contribution rate c_0 (constant in our previous model) has been modified to $c_0 e^{\alpha t}$. This now allows the annual contribution rate to be continuously compounded over a time period t and at an average annual growth rate of α. The above equation is yet another very nice example of a solvable Bernoulli-in-form differential equation. As a review, we shall do so in the next example.

299

Ex 8.15: Solve $\dfrac{dP}{dt} = rP + c_0 e^{\alpha t}$: $P(0) = P_0$.

$\overset{1}{\mapsto} : \dfrac{dP}{dt} = rP + c_0 e^{\alpha t} \Rightarrow$

$\dfrac{dP}{dt} - rP = c_0 e^{\alpha t}$

$\overset{2}{\mapsto}$: Form the integrating factor e^{-rt}

$\overset{3}{\mapsto} : \dfrac{dP}{dt} = rP + c_0 e^{\alpha t} \Rightarrow$

$e^{-rt}\left(\dfrac{dP}{dt} - rP\right) = c_0 e^{-rt} \cdot e^{\alpha t} \Rightarrow$

$\dfrac{d}{dt}(Pe^{-rt}) = c_0 e^{(\alpha - r)t} \Rightarrow$

$Pe^{-rt} = \int c_0 e^{(\alpha - r)t}\, dt + C \Rightarrow$

$Pe^{-rt} = \dfrac{c_0}{\alpha - r} \cdot e^{(\alpha - r)t} + C \Rightarrow$

$P = P(t) = \dfrac{c_0}{\alpha - r} \cdot e^{\alpha t} + Ce^{rt}$

$\overset{4}{\mapsto}$: Apply $P(0) = P_0$

$P(0) = P_0 \Rightarrow$

$P_0 = \dfrac{c_0}{\alpha - r} \cdot e^{\alpha(0)} + Ce^{r(0)} \Rightarrow$

$P_0 = \dfrac{c_0}{\alpha - r} + C \Rightarrow C = P_0 - \dfrac{c_0}{\alpha - r} \Rightarrow$

$P(t) = P_0 e^{rt} + \dfrac{c_0}{r - \alpha}\left[e^{rt} - e^{\alpha t}\right] \therefore$

Ex 8.16: Repeat **Ex 8.14** letting the annual contribution rate continuously compound with $\alpha = 3\%$.

A) $\overset{1}{\mapsto}$: No change $P(5) = \$10,000.00e^{0.06(5)} = \$13,498.58$.

$\overset{2}{\mapsto}$: The output from Stage 1 is still input to Stage 2. But now, we have an additional growth mechanism acting for 38 years.

$$P(38) = 13,498.58e^{0.08(38)} + \frac{3000}{0.08 - .03}(e^{0.08(38)} - e^{0.03(38)}) \Rightarrow$$

$$P(38) = \$148,797.22 + \$1,066,708.49 \Rightarrow$$

$$P(38) = \$1,215,500.71$$

B) The % of the final total accrued by the initial $\$10,000.00$ is

$$\frac{\$148,792.22}{\$1,215,500.71} = .122 = 12.2\%$$

C) $PV = \$1,215,500.71e^{-0.03(43)} = \$334,591.83$.

The final annual contribution is $\$3000.00e^{0.03(38)} = \$9380.31$. You should note that the total contribution throughout the 38 years is given by

$$\int_{0}^{38} \$3000.00e^{0.03t}\, dt = \$100,000.00e^{0.03t} \Big|_{0}^{38} = \$212,676.83.$$

Most of us don't receive a large amount of money early in our lives. That is the reason we are a nation primarily made up of middle-class individuals. So with this in mind, we will forgo the early inheritance in our next example.

Ex 8.17: Assume we start our investment program at age 25 with an annual contribution of $\$3000.00$ grown at a rate of $\alpha = 5\%$ per year. Also assume an aggressive annual interest rate of $r = 10\%$ (experts tell us that this is still doable in the long term through smart investing). How much is our nest egg worth at age 68 in today's terms assuming a 3% annual inflation rate?

$$\overset{1}{\mapsto} : P(43) = \frac{3000}{0.10 - 0.05}(e^{0.10(43)} - e^{0.05(43)}) \Rightarrow$$

$$P(43) = \$3,906,896.11$$

$$\overset{2}{\mapsto} : PV = \$3,906,896.11e^{-0.03(43)} = \$1,075,454.35$$

Note: With work and perseverance, we have finally achieved our million dollars.

To close our discussion, there is no end to the investment models that one can make. For example, we can further alter

$$\frac{dP}{dt} = rP + c_0 e^{\alpha t} : P(0) = P_0$$

by writing

$$\frac{dP}{dt} = rP + c_0(1 + \beta t)e^{\alpha t} : P(0) = P_0.$$

This last expression reflects both planned growth of our annual contribution according to the new parameter β and the "automatic" continuous growth due to salary increases (cost-of-living, promotions, etc.). Uncontrollable Interest rates are usually left fixed and averages used throughout the projection period. Finally, projection periods can be broken up into sub-periods (or stages) when major changes occur. In such cases, the analysis also proceeds in the same way: the output from the current stage becomes the input for the next stage.

Note: An economic change in my lifetime is that a person's retirement is becoming more a matter of individual responsibility and less a matter of government or corporate responsibility.

The Way to Economic Security

You must first plan smart. Then, you must do smart!

Calculus can only be of help in the former!

8.3.3) Installments—Loans and Annuities

Loans and annuities are in actuality investment plans in reverse. In either case, one starts with a given amount of money and chips away at the principle over time until the principle is depleted. The governing differential equation for either case is

---

Differential Equation for Continuous Principle Reduction

$$\frac{dP}{dt} = rP - c_0 : P(0) = P_0 \Rightarrow$$

$$P(t) = P_0 e^{rt} - \frac{c_0}{r}(e^{rt} - 1)$$

Instead of being an annual contribution rate, the parameter c_0 is now an annual payout or payment rate. Of particular concern is the annual payment/payout needed to amortize the loan/annuity over a period of T years. To determine this, set $P(T) = 0$ and solve for the corresponding c_0:

$$P(T) = P_0 e^{rT} - \frac{c_0}{r}(e^{rT} - 1) = 0 \Rightarrow$$

$$c_0 = A_{nnual} = \frac{rP_0 e^{rT}}{e^{rT} - 1} \Rightarrow$$

$$M_{onthly} = \frac{rP_0 e^{rT}}{12(e^{rT} - 1)}$$

The parameter c_0 is the annual payment (or payout) needed to amortize the initial principle P_0 over a period of T years at a fixed interest rate r. It is easily converted to monthly payments by dividing by 12. The above formula is based on continuous reduction of principle whereas in practice, principle reduction typically occurs twelve times a year.

303

But as with the case for the continuous principle growth model, the continuous principle reduction model works extremely well when the number of compounding or principle recalculation periods exceeds four per year. Below are four payment/payout formulas based on this model that we will leave to the reader to verify.

Continuous Interest Mortgage/Annuity Formulas

1) First Month's Interest: $I_{1st} = \dfrac{rP_0}{12}$

2) Monthly Payment: $M = \dfrac{rP_0 e^{rT}}{12(e^{rT} - 1)}$

3) Total Interest (I) Repayment : $I = P_0 \left[\dfrac{rTe^{rT}}{e^{rT} - 1} - 1 \right]$

4) Total Repayment ($A = P_0 + I$): $A = \dfrac{rTP_0 e^{rT}}{e^{rT} - 1}$

Ex 8.18: You borrow $\$250,000.00$ for 30 years at an interest rate of 5.75%. Calculate the monthly payment, total repayment , and total interest repayment assuming no early payout.

$\overset{1}{\mapsto}:\ M = \dfrac{0.0575(\$250,000.00)e^{0.0575(30)}}{12(e^{0.0575(30)} - 1)} = \$1457.62$

$\overset{2}{\mapsto}:\ A = \dfrac{0.0575(30)(\$250,000.00)e^{0.0575(30)}}{(e^{0.0575(30)} - 1)} = \$524,745.50$

$\overset{3}{\mapsto}:\ I = A - P_0 = \$524,745.50 - \$250,000.00 = \$274,745.51$

You have probably heard people say, I am paying my mortgage off in cheaper dollars.

This statement refers to the effects of inflation on future mortgage payments. Future mortgage payments are simply not worth as much in today's terms as current mortgage payments. In fact, if we project t years into the loan and the annual inflation rate has been i throughout that time period, then the present value of our future payment is

$$M_{PV} = \frac{rP_0 e^{rT}}{12(e^{rT} - 1)} e^{-it}.$$

To illustrate, in **Ex 8.18** the present value of a payment made 21 years from now (assuming a 3% annual inflation rate) is

$$M_{PV} = \$1457.62 e^{-0.03(21)} = \$776.31.$$

Thus, under stable economic conditions, our ability to comfortably *afford* the mortgage should increase over time. This is a case where inflation works in our favor.

Continuing with this discussion, if we are paying off our mortgage with cheaper dollars, then what is the present value of the total repayment? A simple definite integral—interpreted as continuous summing—provides the answer

---

Present Value of Total Mortgage Repayment

$$A_{PV} = \int_0^T \left[\frac{rP_0 e^{rT}}{e^{rT} - 1} \right] e^{-it} dt \Rightarrow$$

$$A_{PV} = \frac{\left(\frac{r}{i}\right) P_0 (e^{rT} - e^{(r-i)T})}{(e^{rT} - 1)}$$

In **Ex 8.18**, the present value of the total 30-year repayment stream is $A_{PV} = \$345,999.90$.

Ex 8.19: Compare M, A, and A_{PV} for a mortgage with $P_0 = \$300,000.00$ if the interest rates are: $r_{30} = 6\%$, $r_{20} = 5.75\%$, and $r_{15} = 5.0\%$. Assume a steady annual inflation rate of $i = 3\%$ and no early payout.

In this example, we dispense with the calculations (the reader should check) and present the results in **Table 8.2**.

Fixed Rate Mortgage with $P_0 = \$300,000.00$				
Terms	r	M	A	A_{PV}
$T = 30$	6.00%	$1797.05	$646,938.00	$426,569.60
$T = 20$	5.75%	$2103.57	$504,856.80	$379.642.52
$T = 15$	5.00%	$2369.09	$426,436.20	$343,396.61

Table 8.2: Fixed Rate Mortgage Comparison

Table 8.2 definitely shows the mixed advantages/disadvantages of choosing a short-term or long-term mortgage. For a fixed principle, long-term mortgages have lower monthly payments. They also have a much higher overall repayment, although the total repayment is dramatically reduced by the inflation factor. The mortgage decision is very much an individual one and should be done considering all the facts within the scope of the broader economic picture.

Our last example in the book is a simple annuity problem.

Ex 8.20: You finally retire at age 68 and invest the hard-earned money via **Ex 8.17** in an annuity paying 4.5% to be amortized by age 90. What is the monthly payment to you in today's terms?

Annuities are simply mortgages in reverse. Payouts are made *instead of payins* until the principle is reduced to zero.

Continuing, the phrase, in today's terms, means we let $P_0 = PV$.

$$\overset{1}{\mapsto} : PV = P_0 = \$1,075,454.35$$

$$\overset{2}{\mapsto} : M = \frac{(0.045)(\$1,075,454.35)e^{(0.045)24}}{12(e^{(0.045)24} - 1)} \Rightarrow$$

$$M = \$6,106.79$$

The monthly income provided by the annuity looks very reasonable referencing to the year 2003. But, unfortunately, it is a fixed-income annuity that will continue as fixed for 24 years. And, what happens during that time? Inflation! To calculate the value of that monthly payment (in today's terms) at age 84, our now well-known inflation factor is used to obtain

$$M = \$6,106.79 e^{-.03(16)} = \$3778.80.$$

To close, the power provided by the techniques in this short section on finance is nothing short of miraculous. We have used Bernoulli-in-form differential equations to model and solve problems in inflation, investment planning, and installment payment determination (whether loans or annuities). We have also revised the interpretation of the definite integral as a continuous sum in order to obtain the present value of a total repayment stream many years into the future. These economic and personal issues are very much today's issues, and calculus still very much remains a worthwhile tool-of-choice (even for mundane earthbound problems) some 300 years after its inception.

$$\int_a^b \overset{\bullet\bullet}{\cup} dx \int_a^b \overset{\bullet\bullet}{\cup} dx$$

Chapter/Section Exercises

1) Fill in the following table assuming an average annual inflation rate of $i = 3\%$

Fixed Rate Mortgage with $P_0 = \$230{,}000.00$				
Terms	r	M	A	A_{PV}
$T = 30$	6.50%			
$T = 20$	6.25%			
$T = 15$	5.50%			

2) A) Assuming continuous 8% interest, how long will it take to quadruple an initial investment of $\$10{,}000.00$ B) What continuous interest rate would one need to triple this same investment in ten years? C) Increase it twenty times in 40 years?

3) You start an Individual Retirement Account (IRA) at age 25 by investing $\$7000.00$ per year in a very aggressive growth fund having an annual rate of return that averages 13%. Five years later, you roll over the proceeds from this fund into a blue-chip growth fund whose average long-term-rate-of-return is 9% annually. During the first year, you continue with the $\$7000.00$ annual contribution. After that, you increment your annual contribution by 5% via the model $\$7000.00e^{0.05t}$ ceasing contributions at age 69.

A) Assuming continuous and steady interest rates, project the face value of your total investment when you reach age 69.

B) What is the present value of total projected in part A) if inflation holds at a steady rate of $i = 3\%$ throughout the 44-year period?

C) What is the present value of the monthly payment associated with an annuity bought at age 69 with the total in B). Assume the annuity pays a fixed 4% and is amortized at age 100.

D) If you actually lived to be age 100, what would be the present value of the final annuity payment if the inflation rate remains relatively constant at 3% throughout this 75-year period?

9) Magnificent Shoulders

"Let us now praise the worthy
Whose minds begat us."
The Apocrypha (paraphrase)

When I was a small child, my father and mother took me on a summer driving trip to California. During the roundabout return, they stopped to see Mount Rushmore in South Dakota. I have no real memories of Mount Rushmore, just a few old black-and-white photographs that remind me that I was once there, there as a five-old old in the summer of 1953. Four great images stared down from that granite mountain fifty years ago as they still do today. The only difference is that I can feel today what it means. Then, I was clueless.

In similar fashion, the science of calculus has five great personages associated with its development. Their names are carved on the mountain in **Figure 9.1**.

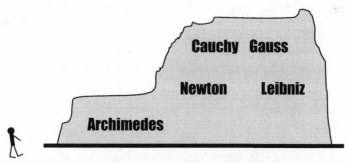

Figure 9.1: The Mount Rushmore of Calculus—
Higher Names are More Recent

Three of these personages happen to be the three greatest mathematicians of all history. It seems that the subject of calculus has a way of attracting the brightest and the best.

However, we must not forget the hundreds of others who helped build the Mount Rushmore of Calculus—*not only a great achievement in its own right but also an enabler of many other great achievements in Western Science.* I have introduced just a few of the calculus builders, such as Bernoulli, within these pages.

And, calculus is still developing. More importantly, it still has a significant role to play, even in a computer age, as this discipline is utilized on a daily basis to formulate equations governing all sorts of physical phenomena. And, what type of equations? By now we should have no trouble citing the answer: differential equations. The wee x —as I first amusingly referred to it in Chapter 4—is here to stay. Turning to how calculus is still developing, we will end our collective adventure with a single example. But first, we need to provide two pages of background on the topic of *partial derivatives*.

Throughout this book, we have restricted our study to functions of a single independent variable and their associated derivatives called ordinary derivatives. We have also examined in brief the associated differential equations called ordinary differential equations (Chapter 8). *Not studied in this book* are functions of *two or more independent variables*, which also have various derivatives, called *partial derivatives*. The associated differential equations that are formulated using partial derivatives are called *partial differential equations*. Using a slight process change, *partial derivatives* can be *taken* using the same differentiation rules as presented in Chapter 5. For example, let

$$f(x, y) = xy^2$$

be a function of the two independent variables x and y . Then one could take a *partial derivative* or 'part of a derivative' with respect to x by treating y as a momentary constant to obtain

$$\frac{\partial f}{\partial x} = y^2 .$$

The symbol '∂' is used to denote partial differentiation and functions akin to the symbol 'd' in ordinary differentiation.

Likewise, a complete rendering of the symbol $\dfrac{\partial f}{\partial x}$ is 'the partial derivative of f with respect to the independent variable x'. For the function $f(x, y) = xy^2$, several other partial derivatives are also obtainable.

$$\frac{\partial f}{\partial y} = 2xy$$

$$\frac{\partial^2 f}{\partial x^2} = 0$$

$$\frac{\partial^2 f}{\partial y^2} = 2x$$

$$\frac{\partial^2 f}{\partial x \partial y} = 2y$$

$$\frac{\partial^2 f}{\partial y \partial x} = 2y$$

$$\frac{\partial^3 f}{\partial x^3} = 0$$

$$\frac{\partial^3 f}{\partial y^3} = 0$$

In each of the seven examples above, the partial derivative is calculated by momentarily holding constant the unnamed independent variable and using ordinary differentiation rules with respect to the named independent variable. For the case

$$\frac{\partial^2 f}{\partial x \partial y} = 2y,\text{ called a }\textit{mixed partial derivative,}$$

one first differentiates with respect to y and then with respect to x as indicated by the symbol $\partial x \partial y$, keeping the standard 'inward-first' principle of algebra in mind.

An interesting result is the equality of mixed partials

$$\frac{\partial^2 f}{\partial x \partial y} = \frac{\partial^2 f}{\partial y \partial x},$$

a result general true for 'well-behaved' functions $f(x, y)$. An example of one partial differential equation combining the partial derivatives just calculated is

$$\frac{\partial^2 f}{\partial x^2} + \frac{\partial^2 f}{\partial y^2} = 2x.$$

Another is

$$\frac{\partial^2 f}{\partial y^2} + \frac{\partial f}{\partial y} = 2x(1 + y).$$

To solve a partial differential equation (e.g. the ones above) or partial differential system of equations—*several partial differential equations coupled in unison in order to describe a complex phenomena*—means to find the function or a group of functions that satisfies either a single equation or system of equations. Partial differential equations normally come with an associated set of boundary or initial conditions, perhaps

$$f(2,1) = 9 \quad \text{along with} \quad \frac{\partial f(1,0)}{\partial y} = 5,$$

all of which must be built into the solution framework.

      With this brief background, we are now ready to present our example. One of the Millennial Problems (seven problems where each carries a cash prize of one-million dollars if solved) is to find a general solution for the complete set of Navier-Stokes Equations. The Navier-Stokes Equations are a coupled set of *six partial differential equations* governing mass, motion, and energy transfer in a general fluidic flow field having six dependent variables: density ρ, pressure p, three velocity components u, v, w, and energy q.

Each physical quantity mathematically linked within the Navier-Stokes partial differential system is a function of the three independent variables x, y, z of ordinary 3D space and the one independent variable t of time. Therefore, we have

$$\rho = \rho(x, y, z, t)$$
$$p = p(x, y, z, t)$$
$$u = u(x, y, z, t)$$
$$v = v(x, y, z, t)$$
$$w = w(x, y, z, t)$$
$$q = q(x, y, z, t)$$

Mind-boggling in their complexity, the Navier-Stokes Equations govern the simultaneous behavior of six dependent variables where each dependent variable is a function of four dependent variables. They can be formulated for either a differential fluid element moving through ordinary three-dimensional (3D) space or a fixed differential element of 3D space through which fluid particles move. See **Figure 9.2**. Even more mind-boggling is the fact that the Navier-Stokes Equations can be formulated in their entirety using the differential concept as presented in this book.

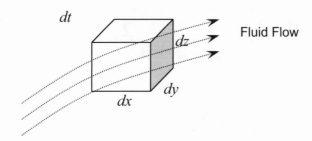

Figure 9.2: A Fixed Differential Element in Space

The most basic equation in the Navier Stokes *coupled partial-differential system* is called the Continuity Equation, which states the requirement for conservation of mass in a dynamic flow field *using differential terms*. We simply display the formulation.

$$\frac{\partial \rho}{\partial t} + \frac{\partial(\rho u)}{\partial x} + \frac{\partial(\rho v)}{\partial y} + \frac{\partial(\rho w)}{\partial z} = 0$$

Navier-Stokes Continuity Equation

In short, the Navier-Stokes equations govern anything that flies, or swims, or sinks below the waves! Restrictive special cases of the Navier-Stokes equations govern both the airflow over the wing of our latest flying machine and the airflow through the core of the modern gas-turbine marvel that powers the same. Even the circulatory systems of the live flesh-and-blood human beings who ride in these flying machines can be described by the same Navier-Stokes equations, even the global weather system through which the flying machines pass on an hourly basis. And, unfortunately, the Navier-Stokes equations have their blind and unchanging way when not adhered to such as in the Columbia disaster in February 2003. Today, fairly complex cases of the general Navier-Stokes Equations can be solved using solved powerful computer methodologies. But, they have not been solved to date in their entirety using classical mathematical methods (including those methods employing limits). Einstein gave up finding a general solution—said it was too difficult. Hence, the quest continues as a new scientific generation seeks to find an all-encompassing million-dollar solution to the Navier-Stokes Equations.

Again, the wonderful irony is that the Navier-Stokes Equations are easily formulated with Leibniz and Newton's 17th Century dx, *a 'bonny wee thing' as Robert Burns might have said—our bonny end to this primer.*

Continue to be challenged!

Figure 9.3: Dreams

O Icarus...

I ride high...
With a whoosh to my back
And no wind to my face,
Folded hands
In quiet rest—
Watching...O Icarus...
The clouds glide by,
Their fields far below
Of gold-illumed snow,
Pale yellow, tranquil moon
To my right—evening sky.

And Wright...O Icarus...
Made it so—
Silvered chariot streaking
On tongues of fire leaping—
And I will soon be sleeping
Above your dreams...

August 2001

100th Anniversary of Powered Flight
1903—2003

$$\int_a^b \overset{\bullet\bullet}{\cup} dx \ \int_a^b \overset{\bullet\bullet}{\cup} dx \ \int_a^b \overset{\bullet\bullet}{\cup} dx$$

Epilogue: Sputnik Plus Fifty

"And so my fellow Americans,
Ask not what your country can do for you;
Ask what you can do for your country. "
        JFK Inaugural Address; January 20, 1961

"I believe this nation should commit itself
To achieving the goal, before this decade is out,
Of landing a man on the moon
And returning him safely to earth."
        JFK, Congressional Address; May 25, 1961

This year America marks the fiftieth anniversary of the world's first artificial satellite, Sputnik I, launched on Friday, 04 October 1957 by the former Soviet Union. I distinctly remember waking up to this news on a bright Saturday morning just 19 days prior to my 10th birthday. Describing my reaction when told by my father when seated at the breakfast table, I would have to say it was one of bewilderment combined with "shock and awe". As a 'Leave-it-to-Beaver' type of kid growing up with the same, I and my neighborhood cohorts thought America was invincible. We 'sat on the knees' of those who helped win War II and thrilled to their stories and all things American. In contrast, we snickered at Japanese-made toys and trinkets passed out at carnivals and school festivals—trophies of war if you will, small harmless tokens offered in homage by a defeated enemy. Even though we grew up in the shadow of The Bomb and played many rounds of "duck and cover" during the early years of the Cold War; overall, the fifties were a heady time for us baby boomers, a time of Lassie-like bliss and American prowess—until 04 October, 1957.

Sputnik I, 184 pounds of beeping Soviet force, circled the globe once every 96 minutes and, with every revolution, reminded the free world that science, mathematics, and the associated technological dominance will ultimately determine who is free and who is not. Several weeks after the launch, I was standing in my backyard with my father and several neighbors waiting for Sputnik to emerge from the western horizon, above the rapidly-disappearing twilight of a clear November evening. Indeed it did, right on schedule.

Figure E.1: November 1957

Sputnik then looped silently and ominously towards the east, steadily blinking via reflected sunlight as it spun on its own axis. Just as ominous was the mood in that backyard, a mood coming from somber and grown men—some of which, as my father, were Air Force. Their whispers were whispers of concern for America and its future, and their visions were those of nuclear-tipped missions aimed at the United States with no counter. Needless to say, one ten-year was able to catch their sense of vulnerability—as opposed to prior invincibility—that November evening. As quickly as it rose, Sputnik sank from view amongst the rising stars within a minute or two. The world had been permanently changed. In response, America got busy and placed more emphasis on mathematics and science education. We as a nation met the Sputnik challenge on two critical fronts, technological (in terms of the arms race) and economic. Almost 32 years later on 9 November 1989, the Berlin Wall came tumbling down, signaling the end of the Cold War.

A victorious bliss settled on America; and, in many ways, we remained in that blissful state throughout the dot.com 90s until 11 September 2001 signaling the start of the War on Terror. Unbeknownst to us, a second war on the global economic front had already started. One could date the start of this war as 02 August 1995, the date of the Netscape IPO. Netscape along with untold millions of miles of light-speed communication cables gave instant access to the best of the best from around the world. Need brains? Got brains—in China, in India, wherever.

This means the bright kid in Calcutta has the same chance of landing a choice job with Procter & Gamble as the bright kid in Cincinnati. And if the kids in Cincinnati aren't that bright—or motivated—guess who wins! When outsourcing was confined to cheap low-tech products on American shelves—clothing, kitchen wares, lawnmowers, etc—we didn't seem to mind, but to expand the outsourcing concept to include engineering and technical expertise coming from hungry and able minds beyond American shores was another matter. As an American who has pursued dual careers in defense and education, I am worried in much the same way that I was worried in 1957. I perceive an emerging threat with the only solution being superior technological prowess as it is *home-grown* on our native shores. The opening scenes of the movie "Gladiator" disturbingly come to mind when Rome, with clear technological superiority, easily conquers a multitude of chest-beating barbarians lulled into thinking they could overcome this sophisticated enemy with manliness alone.

Our nation's math and science teachers are allies with all of us who directly serve the cause of maintaining America's military preeminence in the world. Together, we engaged in a common battle—that of keeping America the greatest human experiment in democratic government ever created on Planet Earth. One of the keys to preserving our government and associated way of life in a dangerous and unstable world is to insure an able and continuing supply of well-trained young citizens who can 1) compete with the best of the best from around the world and 2) insure that our critical freedom-preserving technologies are also the best of the best. Remember that the American experiment is still an experiment. God forbid that it should ever fail! And it will not fail if each of us in whatever capacity—teacher, student, parent, grandparent, mentor—gets busy with the learning and building task at hand.

Appendix A: Formulas from Algebra

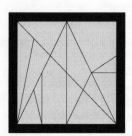

Archimedes' Stomachion: Circa 250BCE

This is the earliest known geometric puzzle

A.1: What is a Variable?

In the fall of 1961, I first encountered the monster called x in my high-school freshman algebra class. The letter x is still a monster to many, whose real nature has been confused by such words as *variable* and *unknown*: perhaps the most horrifying description of x ever invented! Actually, x is very easily understood in terms of a language metaphor. In English, we have both proper nouns and pronouns where both are distinct and different parts of speech. Proper nouns are specific persons, places, or things such as John, Ohio, and Toyota. Pronouns are nonspecific persons or entities such as he, she, or it.

To see how the concept of pronouns and nouns applies to algebra, we first examine arithmetic, which can be thought of as a precise language of quantification having four action verbs, a verb of being, and a plethora of proper nouns. The four action verbs are addition, subtraction, multiplication, and division denoted respectively by $+, -, \cdot, \div$. The verb of being is called *equals* or *is*, denoted by $=$. Specific numbers such as 12, 3.4512, $23\frac{3}{5}$, $\frac{123}{769}$, 0.00045632, -45, , serve as the arithmetical equivalent to proper nouns in English. So, what is x? x is merely a *nonspecific number*, the mathematical equivalent to a pronoun in English. English pronouns greatly expand our capability to describe and inform in a general fashion. Hence, pronouns add increased flexibility to the English language. Likewise, mathematical pronouns—such as x, y, z—greatly expand our capability to quantify in a general fashion by adding flexibility to our language of arithmetic. Arithmetic, with the addition of x, y, z and other mathematical pronouns as a new part of speech, is called algebra.

<u>In Summary</u>: Algebra can be defined as a generalized arithmetic that is much more powerful and flexible than standard arithmetic. The increased capability of algebra over arithmetic is due to the inclusion of the mathematical pronoun x and its associates y, z, etc. A more user-friendly name for *variable* or *unknown* is *pronumber*.

A.2: Field Axioms

The field axioms *decree* the fundamental operating properties of the real number system and provide the basis for all advanced operating properties in mathematics. Let a, b & c be any three real numbers (pronumbers). The field axioms are as follows.

Properties	Addition +	Multiplication ·
Closure	$a + b$ is a unique real number	$a \cdot b$ is a unique real number
Commutative	$a + b = b + a$	$a \cdot b = b \cdot a$
Associative	$(a + b) + c = a + (b + c)$	$(ab)c = a(bc)$
Identity	$0 \Rightarrow a + 0 = a$	$1 \Rightarrow a \cdot 1 = a$
Inverse	$a \Rightarrow a + (-a) = 0$ $\Rightarrow (-a) + a = 0$	$a \neq 0 \Rightarrow a \cdot \dfrac{1}{a} = 1$ $\Rightarrow \dfrac{1}{a} \cdot a = 1$
Distributive *or* *Linking Property*	$a \cdot (b + c) = a \cdot b + a \cdot c$	
Transitivity	$a = b$ & $b = c \Rightarrow a = c$ $a > b$ & $b > c \Rightarrow a > c$ $a < b$ & $b < c \Rightarrow a < c$	
Note: $ab = a(b) = (a)b$ *are alternate representations of* $a \cdot b$		

A.3: Subtraction, Division, Signed Numbers

1. Definitions:

   Subtraction: $\quad a - b \equiv a + (-b)$

   Division: $\qquad a \div b \equiv a \cdot \dfrac{1}{b}$

2. Alternate representation of $a \div b$: $a \div b \equiv \dfrac{a}{b}$

3. Division Properties of Zero

   Zero in numerator: $a \neq 0 \Rightarrow \dfrac{0}{a} = 0$

   Zero in denominator: $\dfrac{a}{0}$ is *undefined*

   Zero in both: $\dfrac{0}{0}$ is *undefined*

4. Demonstration that division-by-zero is *undefined*

   $\dfrac{a}{b} = c \Rightarrow a = b \cdot c$ for all real numbers a

   If $\dfrac{a}{0} = c$, then $a = 0 \cdot c \Rightarrow a = 0$ for all real numbers a,

   an algebraic impossibility

5. Demonstration that attempted division-by-zero leads to erroneous results.

   Let $x = y$; then multiplying both sides by x gives

   $x^2 = xy \Rightarrow x^2 - y^2 = xy - y^2 \Rightarrow$

   $(x - y)(x + y) = y(x - y)$

   Dividing both sides by $x - y$ where $x - y = 0$ gives

   $x + y = y \Rightarrow 2y = y \Rightarrow 2 = 1$.

   The last equality is a false statement.

6. Signed Number Multiplication:

$$(-a) \cdot b = -(a \cdot b)$$
$$a \cdot (-b) = -(a \cdot b)$$
$$(-a) \cdot (-b) = (a \cdot b)$$

7. Table for Multiplication of Signed Numbers: the *italicized words* in the body of the table indicate the resulting sign of the associated product.

Multiplication of $a \cdot b$		
Sign of a	Sign of b	
	Plus	Minus
Plus	*Plus*	*Minus*
Minus	*Minus*	*Plus*

8. Demonstration of the algebraic reasonableness of the laws of multiplication for signed numbers. In both columns, both the middle and rightmost numbers decrease in the expected logical fashion.

$(4) \cdot (5) = 20$	$(-5) \cdot (4) = -20$
$(4) \cdot (4) = 16$	$(-5) \cdot (3) = -15$
$(4) \cdot (3) = 12$	$(-5) \cdot (2) = -10$
$(4) \cdot (2) = 8$	$(-5) \cdot (1) = -5$
$(4) \cdot (1) = 4$	$(-5) \cdot (0) = 0$
$(4) \cdot (0) = 0$	$(-5) \cdot (-1) = 5$
$(4) \cdot (-1) = -4$	$(-5) \cdot (-2) = 10$
$(4) \cdot (-2) = -8$	$(-5) \cdot (-3) = 15$
$(4) \cdot (-3) = -12$	$(-5) \cdot (-4) = 20$
$(4) \cdot (-4) = -16$	$(-5) \cdot (-5) = 25$
$(4) \cdot (-5) = -20$	$(-5) \cdot (-6) = 30$

A.4: Rules for Fractions

Let $\dfrac{a}{b}$ and $\dfrac{c}{d}$ be fractions with $b \neq 0$ and $d \neq 0$.

1. Fractional Equality: $\dfrac{a}{b} = \dfrac{c}{d} \Leftrightarrow ad = bc$

2. Fractional Equivalency: $c \neq 0 \Rightarrow \dfrac{a}{b} = \dfrac{ac}{bc} = \dfrac{ca}{cb}$

3. Addition (like denominators): $\dfrac{a}{b} + \dfrac{c}{b} = \dfrac{a+c}{b}$

4. Addition (unlike denominators):
$$\dfrac{a}{b} + \dfrac{c}{d} = \dfrac{ad}{bd} + \dfrac{cb}{bd} = \dfrac{ad+cb}{bd}$$
    Note: bd is the common denominator

5. Subtraction (like denominators): $\dfrac{a}{b} - \dfrac{c}{b} = \dfrac{a-c}{b}$

6. Subtraction (unlike denominators):
$$\dfrac{a}{b} - \dfrac{c}{d} = \dfrac{ad}{bd} - \dfrac{cb}{bd} = \dfrac{ad-cb}{bd}$$

7. Multiplication: $\dfrac{a}{b} \cdot \dfrac{c}{d} = \dfrac{ac}{bd}$

8. Division: $c \neq 0 \Rightarrow \dfrac{a}{b} \div \dfrac{c}{d} = \dfrac{a}{b} \cdot \dfrac{d}{c} = \dfrac{ad}{bc}$

9. Division (missing quantity): $\dfrac{a}{b} \div c = \dfrac{a}{b} \div \dfrac{c}{1} = \dfrac{a}{b} \cdot \dfrac{1}{c} = \dfrac{a}{bc}$

10. Reduction of Complex Fraction: $\dfrac{\dfrac{a}{b}}{\dfrac{c}{d}} = \dfrac{a}{b} \div \dfrac{c}{d} = \dfrac{ad}{bc}$

11. Placement of Sign: $-\dfrac{a}{b} = \dfrac{-a}{b} = \dfrac{a}{-b}$

A.5: Partial Fractions

Let $P(x)$ be a polynomial expression with degree less than the degree of the factored denominator as shown.

1. Two Distinct Linear Factors:

$$\frac{P(x)}{(x-a)(x-b)} = \frac{A}{x-a} + \frac{B}{x-b}$$

The numerators A, B are given by

$$A = \frac{P(a)}{a-b}, B = \frac{P(b)}{b-a}$$

2. Three Distinct Linear Factors:

$$\frac{P(x)}{(x-a)(x-b)(x-c)} = \frac{A}{x-a} + \frac{B}{x-b} + \frac{C}{x-c}$$

The numerators A, B, C are given by

$$A = \frac{P(a)}{(a-b)(a-c)}, B = \frac{P(b)}{(b-a)(b-c)},$$

$$C = \frac{P(c)}{(c-a)(c-b)}$$

3. N Distinct Linear Factors:

$$\frac{P(x)}{\prod_{i=1}^{n}(x-a_i)} = \sum_{i=1}^{n} \frac{A_i}{x-a_i} \text{ with } A_i = \frac{P(a_i)}{\prod_{\substack{j=1 \\ j \neq i}}^{n}(a_i-a_j)}$$

A.6: Rules for Exponents

1. Addition: $a^n a^m = a^{n+m}$

2. Subtraction: $\dfrac{a^n}{a^m} = a^{n-m}$

3. Multiplication: $(a^n)^m = a^{nm}$

4. Distributed over a Simple Product: $(ab)^n = a^n b^n$

5. Distributed over a Complex Product: $(a^m b^p)^n = a^{mn} b^{pn}$

6. Distributed over a Simple Quotient: $\left(\dfrac{a}{b}\right)^n = \dfrac{a^n}{b^n}$

7. Distributed over a Complex Quotient: $\left(\dfrac{a^m}{b^p}\right)^n = \dfrac{a^{mn}}{b^{pn}}$

8. Definition of Negative Exponent: $\dfrac{1}{a^n} \equiv a^{-n}$

9. Definition of Radical Expression: $\sqrt[n]{a} \equiv a^{\frac{1}{n}}$

10. Definition when No Exponent is Present: $a \equiv a^1$

11. Definition of Zero Exponent: $a^0 \equiv 1$

12. Demonstration of the algebraic reasonableness of the definitions for a^0 and a^{-n} via successive divisions by 2. Notice the power decreases by 1 with each division.

$$16 = 32 \div 2 = 2 \cdot 2 \cdot 2 \cdot 2 = 2^4$$
$$8 = 16 \div 2 = 2 \cdot 2 \cdot 2 = 2^3$$
$$4 = 8 \div 2 = 2 \cdot 2 = 2^2$$
$$2 = 4 \div 2 \equiv 2^1$$
$$1 = 2 \div 2 \equiv 2^0$$

$$\tfrac{1}{2} = 1 \div 2 = \tfrac{1}{2^1} \equiv 2^{-1}$$
$$\tfrac{1}{4} = \left[\tfrac{1}{2}\right] \div 2 = \tfrac{1}{2^2} \equiv 2^{-2}$$
$$\tfrac{1}{8} = \left[\tfrac{1}{4}\right] \div 2 = \tfrac{1}{2^3} \equiv 2^{-3}$$
$$\tfrac{1}{16} = \left[\tfrac{1}{8}\right] \div 2 = \tfrac{1}{2^4} \equiv 2^{-4}$$

A.7: Rules for Radicals

1. Basic Definitions: $\sqrt[n]{a} \equiv a^{\frac{1}{n}}$ and $\sqrt[2]{a} \equiv \sqrt{a} \equiv a^{\frac{1}{2}}$

2. Complex Radical: $\sqrt[n]{a^m} = a^{\frac{m}{n}}$

3. Associative: $(\sqrt[n]{a})^m = \sqrt[n]{a^m} = a^{\frac{m}{n}}$

4. Simple Product: $\sqrt[n]{a}\sqrt[n]{b} = \sqrt[n]{ab}$

5. Simple Quotient: $\dfrac{\sqrt[n]{a}}{\sqrt[n]{b}} = \sqrt[n]{\dfrac{a}{b}}$

6. Complex Product: $\sqrt[n]{a}\sqrt[m]{b} = \sqrt[nm]{a^m b^n}$

7. Complex Quotient: $\dfrac{\sqrt[n]{a}}{\sqrt[m]{b}} = \sqrt[nm]{\dfrac{a^m}{b^n}}$

8. Nesting: $\sqrt[n]{\sqrt[m]{a}} = \sqrt[nm]{a}$

9. Rationalizing Numerator for $n > m$: $\dfrac{\sqrt[n]{a^m}}{b} = \dfrac{a}{b\sqrt[n]{a^{n-m}}}$

10. Rationalizing Denominator for $n > m$: $\dfrac{b}{\sqrt[n]{a^m}} = \dfrac{b\sqrt[n]{a^{n-m}}}{a}$

11. Complex Rationalization Process:

$$\dfrac{a}{b+\sqrt{c}} = \dfrac{a(b-\sqrt{c})}{(b+\sqrt{c})(b-\sqrt{c})} \Rightarrow$$

$$\dfrac{a}{b+\sqrt{c}} = \dfrac{a(b-\sqrt{c})}{b^2 - c}$$

Numerator: $\dfrac{a+\sqrt{c}}{b} = \dfrac{a^2 - c}{b(a-\sqrt{c})}$

12. Definition of Surd Pairs: If $a \pm \sqrt{b}$ is a radical expression, then the associated surd is given by $a \mp \sqrt{b}$.

A.8: Factor Formulas

1. Simple Common Factor: $ab + ac = a(b+c) = (b+c)a$
2. Grouped Common Factor:
$$ab + ac + db + dc =$$
$$(b+c)a + d(b+c) =$$
$$(b+c)a + (b+c)d =$$
$$(b+c)(a+d)$$
3. Difference of Squares: $a^2 - b^2 = (a+b)(a-b)$
4. Expanded Difference of Squares:
$$(a+b)^2 - c^2 = (a+b+c)(a+b-c)$$
5. Sum of Squares: $a^2 + b^2 = (a+bi)(a-bi)$ i *complex*
6. Perfect Square: $a^2 \pm 2ab + b^2 = (a \pm b)^2$
7. General Trinomial:
$$x^2 + (a+b)x + ab =$$
$$(x^2 + ax) + (bx + ab) =$$
$$(x+a)x + (x+a)b =$$
$$(x+a)(x+b)$$
8. Sum of Cubes: $a^3 + b^3 = (a+b)(a^2 - ab + b^2)$
9. Difference of Cubes: $a^3 - b^3 = (a-b)(a^2 + ab + b^2)$
10. Difference of Fourths:
$$a^4 - b^4 = (a^2 - b^2)(a^2 + b^2) \Rightarrow$$
$$a^4 - b^4 = (a-b)(a+b)(a^2 + b^2)$$
11. Power Reduction to an Integer:
$$a^4 + a^2 b^2 + b^4 = (a^2 + ab + b^2)(a^2 - ab + b^2)$$
12. Power Reduction to a Radical:
$$x^2 - a = (x - \sqrt{a})(x + \sqrt{a})$$
13. Power Reduction to an Integer plus a Radical:
$$a^2 + ab + b^2 = (a + \sqrt{ab} + b)(a - \sqrt{ab} + b)$$

A.9: Laws of Equality

Let $A = B$ be an algebraic equality and C, D be any quantities.

1. Addition: $A + C = B + C$
2. Subtraction: $A - C = B - C$
3. Multiplication: $A \cdot C = B \cdot C$
4. Division: $\dfrac{A}{C} = \dfrac{B}{C}$ provided $C \neq 0$
5. Exponent: $A^n = B^n$ provided n is an integer
6. Reciprocal: $\dfrac{1}{A} = \dfrac{1}{B}$ provided $A \neq 0, B \neq 0$

7. Means & Extremes: $\dfrac{C}{A} = \dfrac{D}{B} \Rightarrow CB = AD$ if $A \neq 0, B \neq 0$

8. Zero Product Property: $A \cdot B = 0 \iff A = 0$ or $B = 0$

A.10: Laws of Inequality

Let $A > B$ be an algebraic inequality and C be any quantity.

1. Addition: $A + C > B + C$
2. Subtraction: $A - C > B - C$
3. Multiplication: $\begin{aligned} C > 0 &\Rightarrow A \cdot C > B \cdot C \\ C < 0 &\Rightarrow A \cdot C < B \cdot C \end{aligned}$
4. Division: $\begin{aligned} C > 0 &\Rightarrow \dfrac{A}{C} > \dfrac{B}{C} \\ C < 0 &\Rightarrow \dfrac{A}{C} < \dfrac{B}{C} \end{aligned}$
5. Reciprocal: $\dfrac{1}{A} < \dfrac{1}{B}$ provided $A \neq 0, B \neq 0$

A.11: Order of Operations

Step 1: Perform all power raisings in the order they occur from left to right

Step 2: Perform all multiplications and divisions in the order they occur from left to right

Step 3: Perform all additions and subtractions in the order they occur from left to right

Step 4: If parentheses are present, first perform steps 1 through 3 *on an as-needed basis* within the innermost set of parentheses until a single number is achieved. Then perform steps 1 through 3 (*again, on an as-needed basis*) for the next level of parentheses until all parentheses have been systematically removed.

Step 5: If a fraction bar is present, simultaneously perform steps 1 through 4 for the numerator and denominator, treating each as totally-separate problem until a single number is achieved. Once single numbers have been achieved for both the numerator and the denominator, then a final division can be performed.

A.12: Three Meanings of 'Equals'

1. **Equals** is the mathematical equivalent of the English verb "is", the fundamental verb of being. A simple but subtle use of equals in this fashion is $2 = 2$.

2. **Equals** implies an equivalency of naming in that the same underlying quantity is being named in two different ways. This can be illustrated by the expression $2003 = MMIII$. Here, the two diverse symbols on both sides of the equals sign refer to the same and exact underlying quantity.

3. **Equals** states the product (either intermediate or final) that results from a process or action. For example, in the expression $2 + 2 = 4$, we are adding two numbers on the left-hand side of the equals sign. Here, addition can be viewed as a process or action between the numbers 2 and 2. The result or product from this process or action is the single number 4, which appears on the right-hand side of the equals sign.

A.13: Rules for Logarithms

1. Definition of Logarithm to Base $b > 0$:
   $$y = \log_b x \text{ if and only if } b^y = x$$

2. Logarithm of the Same Base: $\log_b b = 1$

3. Logarithm of One: $\log_b 1 = 0$

4. Logarithm of the Base to a Power: $\log_b b^p = p$

5. Base to the Logarithm: $b^{\log_b p} = p$

6. Notation for Logarithm Base 10: $Log x \equiv \log_{10} x$

7. Notation for Logarithm Base e: $\ln x \equiv \log_e x$

8. Change of Base Formula: $\log_b N = \dfrac{\log_a N}{\log_a b}$

9. Product: $\log_b(MN) = \log_b N + \log_b M$

10. Quotient: $\log_b\left(\dfrac{M}{N}\right) = \log_b M - \log_b N$

11. Power: $\log_b N^p = p \log_b N$

12. Logarithmic Simplification Process

    Let $X = \dfrac{A^n B^m}{C^p}$, then

    $$\log_b(X) = \log_b\left(\frac{A^n B^m}{C^p}\right) \Rightarrow$$
    $$\log_b(X) = \log_b(A^n B^m) - \log_b(C^p) \Rightarrow$$
    $$\log_b(X) = \log_b(A^n) + \log_b(B^m) - \log_b(C^p) \Rightarrow$$
    $$\log_b(X) = n \log_b(A) + m \log_b(B) - p \log_b(C) \therefore$$

Note: The use of logarithms transforms complex algebraic expressions where products become sums, quotients become differences, and exponents become coefficients, making the manipulation of these expressions easier in some instances.

A.14: Complex Numbers

1. Definition of the imaginary unit i: i is defined to be the solution to the equation $x^2 + 1 = 0$.
2. Properties of the imaginary unit i:
$$i^2 + 1 = 0 \Rightarrow i^2 = -1 \Rightarrow i = \sqrt{-1}$$
3. Definition of Complex Number: Numbers of the form $a + bi$ where a, b are real numbers
4. Definition of Complex Conjugate: $\overline{a + bi} = a - bi$
5. Definition of Complex Modulus: $|a + bi| = \sqrt{a^2 + b^2}$
6. Addition: $(a + bi) + (c + di) = (a + c) + (b + d)i$
7. Subtraction: $(a + bi) - (c + di) = (a - c) + (b - d)i$

8. Process of Complex Number Multiplication
$$(a + bi)(c + di) =$$
$$ac + (ad + bc)i + bdi^2 =$$
$$ac + (ad + bc)i + bd(-1)$$
$$ac - bd + (ad + bc)i$$

9. Process of Complex Number Division
$$\frac{a + bi}{c + di} =$$
$$\frac{(a + bi)\overline{(c + di)}}{(c + di)\overline{(c + di)}} =$$
$$\frac{(a + bi)(c - di)}{(c + di)(c - di)} =$$
$$\frac{(ac + bd) + (bc - ad)i}{c^2 - d^2} =$$
$$\frac{ac + bd}{c^2 - d^2} + \left(\frac{bc - ad}{c^2 - d^2}\right)i$$

A.15: Quadratic Equations & Functions

1. Quadratic Equation Definition: $ax^2 + bx + c = 0 : a \neq 0$

2. Quadratic Formula with Development:

$$ax^2 + bx + c = 0 \Rightarrow x^2 + \left(\frac{b}{a}\right)x = -\frac{c}{a} \Rightarrow$$

$$x^2 + \left(\frac{b}{a}\right)x + \frac{b^2}{4a^2} = -\frac{c}{a} + \frac{b^2}{4a^2} \Rightarrow \left[x + \left(\frac{b}{2a}\right)\right]^2 = \frac{b^2 - 4ac}{4a^2} \Rightarrow$$

$$x + \left(\frac{b}{2a}\right) = \pm\frac{\sqrt{b^2 - 4ac}}{2a} \Rightarrow x = \frac{-b \pm \sqrt{b^2 - 4ac}}{2a} \quad \therefore$$

3. Solution Discriminator: $b^2 - 4ac$

$b^2 - 4ac > 0 \Rightarrow$ two real solutions

$b^2 - 4ac = 0 \Rightarrow$ one real solution of multiplicity two

$b^2 - 4ac < 0 \Rightarrow$ two complex (conjugates) solutions

4. Solution when $a = 0 \,\& \, b \neq 0$:

$$bx + c = 0 \Rightarrow x = \frac{-c}{b}$$

5. Quadratic-in-Form Equation: $aU^2 + bU + c = 0$ where U is an algebraic expression of varying complexity.

6. Definition of Quadratic Function:

$$f(x) = ax^2 + bx + c = a\left(x + \frac{b}{2a}\right)^2 - \frac{b^2 - 4ac}{4a}$$

7. Axis of Symmetry for Quadratic Function: $x = \frac{-b}{2a}$

8. Vertex for Quadratic Function: $\left(\frac{-b}{2a}, \frac{4ac - b^2}{4a}\right)$

A.16: Cardano's Cubic Solution

Let $ax^3 + bx^2 + cx + d = 0$ be a cubic equation written in standard form with $a \neq 0$

Step 1: Set $x = y - \dfrac{b}{3a}$. After this substitution, the above cubic

becomes $y^3 + py + q = 0$ where $p = \left[\dfrac{c}{a} - \dfrac{b^2}{3a^2} \right]$ and

$$q = \left[\frac{2b^2}{27a^3} - \frac{bc}{3a^2} + \frac{d}{a} \right]$$

Step 2: Define u & v such that $y = u - v$ and $p = 3uv$

Step 3: Substitute for y & p in the equation $y^3 + py + q = 0$.

This leads to $(u^3)^2 + qu^3 - \dfrac{p^3}{27} = 0$, which is quadratic-

in-form in u^3.

Step 4: Use the quadratic formula **1.19.3** to solve for u^3

$$u^3 = \frac{-q + \sqrt{q^2 + \frac{4}{27} p^3}}{2}$$

Step 5: Solve for u & v where $v = \dfrac{p}{3u}$ to obtain

$$u = \sqrt[3]{\frac{-q + \sqrt{q^2 + \frac{4}{27} p^3}}{2}} \quad \& \quad v = -\sqrt[3]{\frac{-q - \sqrt{q^2 + \frac{4}{27} p^3}}{2}}$$

Step 6: Solve for x where $x = y - \dfrac{b}{3a} \Rightarrow x = u - v - \dfrac{b}{3a}$

A.17: Theory of Polynomial Equations

Let $P(x) = a_n x^n + a_{n-1} x^{n-1} + ... + a_2 x^2 + a_1 x + a_0$ be a polynomial written in standard form.

The Eight Basic Theorems

1. Fundamental Theorem of Algebra: Every polynomial $P(x)$ of degree $N \geq 1$ has at least one solution x_0 for which $P(x_0) = 0$. This solution may be real or complex (i.e. has the form $a + bi$).

2. Numbers Theorem for Roots and Turning Points: If $P(x)$ is a polynomial of degree N, then the equation $P(x) = 0$ has up to N real solutions or roots. The equation $P(x) = 0$ has exactly N roots if one counts complex solutions of the form $a + bi$. Lastly, the graph of $P(x)$ will have up to $N-1$ turning points (which includes both relative maxima and minima).

3. Real Root Theorem: If $P(x)$ is of odd degree having all real coefficients, then $P(x)$ has at least one real root.

4. Rational Root Theorem: If $P(x)$ has all integer coefficients, then any rational roots for the equation $P(x) = 0$ must have the form $\frac{p}{q}$ where p is a factor of the constant coefficient a_0 and q is a factor of the lead coefficient a_n. This result is used to form a rational-root possibility list.

5. Complex Conjugate Pair Root Theorem: Suppose $P(x)$ has all real coefficients. If $a + bi$ is a root for $P(x)$ with $P(a + bi) = 0$, then $P(a - bi) = 0$.

6. Irrational Surd Pair Root Theorem: Suppose $P(x)$ has all rational coefficients. If $a + \sqrt{b}$ is a root for $P(x)$ with $P(a + \sqrt{b}) = 0$, then $P(a - \sqrt{b}) = 0$.

7. Remainder Theorem: If $P(x)$ is divided by $(x - c)$, then the remainder R is equal to $P(c)$. This result is used extensively to evaluate a given polynomial $P(x)$ for various x values.

8. Factor Theorem: If c is any number with $P(c) = 0$, then $(x - c)$ is a factor of $P(x)$. This means $P(x) = (x - c) \cdot Q(x)$ where $Q(x)$ is a new, reduced polynomial having degree one less than $P(x)$. The converse
$$P(x) = (x - c) \cdot Q(x) \Rightarrow P(c) = 0 \text{ is also true.}$$

The Four Advanced Theorems

1. Root Location Theorem: Let (a, b) be an interval on the x axis with $P(a) \cdot P(b) < 0$. Then there is a value $x_0 \in (a, b)$ such that $P(x_0) = 0$.

2. Root Bounding Theorem: Divide $P(x)$ by $(x - d)$ to obtain $P(x) = (x - d) \cdot Q(x) + R$. Case $d > 0$: If both R and all the coefficients of $Q(x)$ are positive, then $P(x)$ has no root $x_0 > d$. Case $d < 0$: If the roots of $Q(x)$ alternate in sign—with the remainder R "in sync" at the end—then $P(x)$ has no root $x_0 < d$. Note: Coefficients of zero can be counted either as positive or negative—which ever way helps in the subsequent determination.

3. Descartes' Rule of Signs: Arrange $P(x)$ in standard order as shown in the title bar. The number of positive real solutions equals the number of coefficient sign variations or that number decreased by an even number. Likewise, the number of negative real solutions equals the number of coefficient sign variations in $P(-x)$ or that number decreased by an even number.

4. Turning Point Theorem: Let a polynomial $P(x)$ have degree N. Then the number of turning points for a polynomial $P(x)$ can not exceed $N - 1$.

A.18: Determinants and Cramer's Rule

1. Two by Two Determinant Expansion:

$$\begin{vmatrix} a & b \\ c & d \end{vmatrix} = ad - bc$$

2. Three by Three Determinant Expansion:

$$\begin{vmatrix} a & b & c \\ d & e & f \\ g & h & i \end{vmatrix} = a\begin{vmatrix} e & f \\ h & i \end{vmatrix} - b\begin{vmatrix} d & f \\ g & i \end{vmatrix} + c\begin{vmatrix} d & e \\ g & h \end{vmatrix} =$$

$$a(ei - fh) - b(di - fg) + c(dh - eg) =$$
$$aei - ahf + bfg - bdi + cdh - ceg$$

3. Cramer's Rule for a Two-by-Two Linear System

Given $\begin{aligned} ax + by &= e \\ cx + dy &= f \end{aligned}$ with $D = \begin{vmatrix} a & b \\ c & d \end{vmatrix} \neq 0$

Then $x = \dfrac{\begin{vmatrix} e & b \\ f & d \end{vmatrix}}{D}$ and $y = \dfrac{\begin{vmatrix} a & e \\ c & f \end{vmatrix}}{D}$

4. Cramer's Rule for a Three-by-Three Linear System

Given $\begin{aligned} ax + by + cz &= j \\ dx + ey + fz &= k \\ gx + hy + iz &= l \end{aligned}$ with $D = \begin{vmatrix} a & b & c \\ d & e & f \\ g & h & i \end{vmatrix} \neq 0$

Then $x = \dfrac{\begin{vmatrix} j & b & c \\ k & e & f \\ l & h & i \end{vmatrix}}{D}, y = \dfrac{\begin{vmatrix} a & j & c \\ d & k & f \\ g & l & i \end{vmatrix}}{D}, z = \dfrac{\begin{vmatrix} a & b & j \\ d & e & k \\ g & h & l \end{vmatrix}}{D}$

5. Solution Types in $x_i = \dfrac{Dx_i}{D}$

$$Dx_i = 0, D \neq 0 \Rightarrow x_i = 0$$

$$Dx_i = 0, D = 0 \Rightarrow x_i \text{ has infinite solutions}$$

$$Dx_i \neq 0, D \neq 0 \Rightarrow x_i \text{ has a unique solution}$$

$$Dx_i \neq 0, D = 0 \Rightarrow x_i \text{ has no solution}$$

A.19 Binomial Theorem

Let n and r be positive integers with $n \geq r$.

1. Definition of $n!$: $n! = n(n-1)(n-2)...1$,
2. Special Factorials: $0! = 1$ and $1! = 1$
3. Combinatorial Symbol: $\dbinom{n}{r} = \dfrac{n!}{r!(n-r)!}$
4. Summation Symbols:

$$\sum_{i=0}^{n} a_i = a_0 + a_1 + a_2 + a_3 + a_4 + ... + a_n$$

$$\sum_{i=k}^{n} a_i = a_k + a_{k+1} + a_{k+2} + a_{k+3} ... + a_n$$

5. Binomial Theorem: $(a+b)^n = \displaystyle\sum_{i=0}^{n} \binom{n}{i} a^{n-i} b^i$

6. Sum of Binomial Coefficients when $a = b = 1$:

$$\sum_{i=0}^{n} \binom{n}{i} 1^{n-i} 1^i = (1+1)^n = 2^n$$

7. Formula for the $(r+1)th$ Term: $\dbinom{n}{r} a^{n-r} b^r$

A.20 Arithmetic Series

1. Definition: $S = \sum_{i=0}^{n}(a+ib)$ where b is the common increment

2. Summation Formula for S: $S = \dfrac{(n+1)}{2}[2a+nb]$

A.21 Geometric Series

1. Definition: $G = \sum_{i=0}^{n} ar^i$ where r is the common ratio

2. Summation Formula for G:

$$G = \sum_{i=0}^{n} ar^i \Rightarrow rG = \sum_{i=0}^{n} ar^{i+1} \Rightarrow$$

$$G - rG = \sum_{i=0}^{n} ar^i - \sum_{i=0}^{n} ar^{i+1} = a - ar^{i+1} \Rightarrow$$

$$G = \frac{a(1-r^{i+1})}{1-r}$$

3. Infinite Sum Provided $0 < r < 1$: $\sum_{i=0}^{\infty} ar^i = \dfrac{a}{1-r}$

A.22: Variation or Proportionality Formulas

1. Direct: $y = kx$ 2. Inverse: $y = \dfrac{k}{x}$

3. Joint: $z = kxy$ 4. Inverse Joint: $z = \dfrac{kx}{y}$

5. Direct Power: $y = kx^n$ 6. Inverse Power: $y = \dfrac{k}{x^n}$

$$\int_{a}^{b} \overset{\cdot\cdot}{\cup} dx$$

Appendix B: Formulas from Geometry

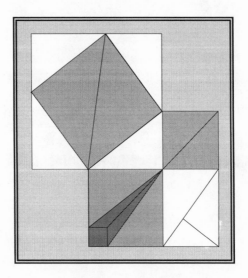

Pythagorean Dreams

Twenty-five centuries of Pythagorean diagrams
Used in various proofs are captured in this one figure.

B.1: The Parallel Postulates

1. Let a point reside outside a given line. Then there is exactly one line passing through the point parallel to the given line.
2. Let a point reside outside a given line. Then there is exactly one line passing through the point perpendicular to the given line.
3. Two lines both parallel to a third line are parallel to each other.
4. If a transverse line intersects two parallel lines, then corresponding angles in the figures so formed are congruent.
5. If a transverse line intersects two lines and makes congruent, corresponding angles in the figures so formed, then the two original lines are parallel.

B.2: Angles and Lines

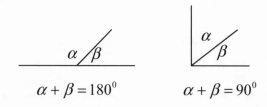

$$\alpha + \beta = 180^{0} \qquad \alpha + \beta = 90^{0}$$

1. Complimentary Angles: Two angles α, β with $\alpha + \beta = 90^{0}$.
2. Supplementary Angles: Two angles α, β with $\alpha + \beta = 180^{0}$
3. Linear Sum of Angles: The sum of the two angles α, β formed when a straight line is intersected by a line segment is equal to 180^{0}
4. Acute Angle: An angle less than 90^{0}
5. Right Angle: An angle exactly equal to 90^{0}
6. Obtuse Angle: An angle greater than 90^{0}

B.3: Triangles

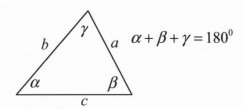

$$\alpha + \beta + \gamma = 180^0$$

1. Triangular Sum of Angles: The sum of the three interior angles α, β, γ in any triangle is equal to 180^0
2. Acute Triangle: A triangle where all three interior angles α, β, γ are acute
3. Right Triangle: A triangle where one interior angle from the triad α, β, γ is equal to 90^0
4. Obtuse Triangle: A triangle where one interior angle from the triad α, β, γ is greater than 90^0
5. Scalene Triangle: A triangle where no two of the three side-lengths a, b, c are equal to another
6. Isosceles Triangle: A triangle where exactly two of the side-lengths a, b, c are equal to each other
7. Equilateral Triangle: A triangle where all three side-lengths a, b, c are identical $a = b = c$ or all three angles α, β, γ are equal with $\alpha = \beta = \gamma = 60^0$
8. Congruent Triangles: Two triangles are congruent (equal) if they have identical interior angles and side-lengths
9. Similar Triangles: Two triangles are similar if they have identical interior angles
10. Included Angle: The angle that is between two given sides
11. Opposite Angle: The angle opposite a given side
12. Included Side: The side that is between two given angles
13. Opposite Side: The side opposite a given angle

B.4 Congruent Triangles

Given the congruent two triangles as shown below

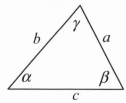 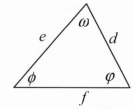

1. Side-Angle-Side (**SAS**): If any two side-lengths and the included angle are identical, then the two triangles are congruent.

2. Angle-Side-Angle (**ASA**): If any two angles and the included side are identical, then the two triangles are congruent.

3. Side-Side-Side (**SSS**): If the three side-lengths are identical, then the triangles are congruent.

4. Three Attributes Identical: If any three attributes—side-lengths and angles—are equal with at least one attribute being a side-length, then the two triangles are congruent. These other cases are of the form Angle-Angle-Side (**AAS**) or Side-Side-Angle (**SSA**).

B.5: Similar Triangles

Given the two similar triangles as shown below

 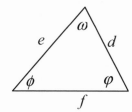

1. Minimal Condition for Similarity: If any two angles are identical (**AA**), then the triangles are similar.

2. Ratio laws for Similar Triangles: Given similar triangles as shown above, then $\dfrac{b}{e} = \dfrac{c}{f} = \dfrac{a}{d}$

B.6: Planar Figures

A is the planar area, P is the perimeter, n is the number of sides.

1. Degree Sum of Interior Angles in General Polygon:
   $$D = 180^0[n-2]$$

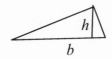

$n = 5 \Rightarrow D = 540^0$

$n = 6 \Rightarrow D = 720^0$

2. Square: $A = s^2 : P = 4s$, s is the length of a side

s

3. Rectangle: $A = bh : P = 2b + 2h$, b & h are the base and height

h

b

4. Triangle: $A = \frac{1}{2}bh$, b & h are the base and altitude

h

b

5. Parallelogram: $A = bh$, b & h are the base and altitude

6. Trapezoid: $A = \frac{1}{2}(B+b)h$, B & b are the two parallel bases and h is the altitude

7. Circle: $A = \pi r^2$: $P = 2\pi r$ where r is the radius, or $P = \pi d$ where $d = 2r$, the diameter.

8. Ellipse: $A = \pi ab$; a & b are the half lengths of the major & minor axes

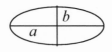

B.7: Solid Figures

A is total surface area, V is the volume

1. Cube: $A = 6s^2$: $V = s^3$, s is the length of a side

2. Sphere: $A = 4\pi r^2 : V = \frac{4}{3}\pi r^3$, r is the radius

3. Cylinder: $A = 2\pi r^2 + 2\pi r l : V = \pi r^2 l$, $r \& l$ are the radius and length

4. Cone: $A = \pi r^2 + 2\pi r t : V = \frac{1}{3}\pi r^2 h$, $r \& t \& h$ are the radius, slant height, and altitude

5. Pyramid (square base): $A = s^2 + 2st : V = \frac{1}{3}s^2 h$, $s \& t \& h$ are the side, slant height, and altitude

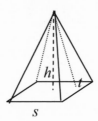

B.8: Pythagorean Theorem

1. Statement: Let a right triangle $\triangle ABC$ have one side $\overline{AC}$ of length x, a second side $\overline{AB}$ of length y, and a hypotenuse (long side) $\overline{BC}$ of length z. Then $z^2 = x^2 + y^2$

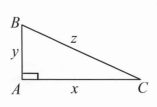
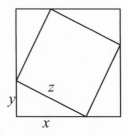

2. Proof: Construct a big square by bringing together four congruent right triangles.

    The area of the big square is given by

    $$A = (x + y)^2 \text{ , or equivalently by}$$

    $$A = z^2 + 4\left(\frac{xy}{2}\right)$$

    Equating:

    $$(x + y)^2 = z^2 + 4\left(\frac{xy}{2}\right) \Rightarrow$$
    $$x^2 + 2xy + y^2 = z^2 + 2xy \Rightarrow$$
    $$x^2 + y^2 = z^2 \Rightarrow$$
    $$z^2 = x^2 + y^2 \ \therefore$$

B.9: Heron's Formula

Let $s = \frac{1}{2}(a+b+c)$ be the semi-perimeter of a general triangle and A be the internal area enclosed by the same.

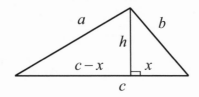

Heron's Formula: $A = \sqrt{s(s-a)(s-b)(s-c)}$

B.10: Distance and Line Formulas

Let (x_1, y_1) and (x_2, y_2) be two points where $x_2 > x_1$.

1. 2-D Distance Formula: $D = \sqrt{(x_2-x_1)^2 + (y_2-y_1)^2}$
2. 3-D Distance Formula: For the points (x_1, y_1, z_1) and (x_2, y_2, z_2), $D = \sqrt{(x_2-x_1)^2 + (y_2-y_1)^2 + (z_2-z_1)^2}$
3. Midpoint Formula: $\left(\dfrac{x_1+x_2}{2}, \dfrac{y_1+y_2}{2} \right)$

Line Formulas

4. Slope of Line: $m = \dfrac{y_2-y_1}{x_2-x_1}$
5. Point/Slope Form: $y-y_1 = m(x-x_1)$
6. General Form: $Ax + By + C = 0$
7. Slope/Intercept Form: $y = mx + b$ where $\left(\dfrac{-b}{m}, 0 \right)$ and $(0, b)$ are the x and y Intercepts:

8. Intercept/Intercept Form: $\dfrac{x}{a} + \dfrac{y}{b} = 1$ where $(a,0)$ and $(0,b)$ are the x and y intercepts

9. Slope Relationship between two Parallel Lines L_1 and L_2 having slopes m_1 and m_2: $m_1 = m_2$

10. Slope Relationship between two Perpendicular Lines L_1 and L_2 having slopes m_1 and m_2: $m_1 = \dfrac{-1}{m_2}$

11. Slope of Line Perpendicular to a Line of Slope m: $\dfrac{-1}{m}$

B.11: Formulas for Conic Sections

1. General: $Ax^2 + Bxy + Cy^2 + Dx + Ey + F = 0$

2. Circle of Radius r Centered at (h,k):
$$(x-h)^2 + (y-k)^2 = r^2$$

3. Ellipse Centered at (h,k): $\dfrac{(x-h)^2}{a^2} + \dfrac{(y-k)^2}{b^2} = 1$

   I) If $a > b$, the two foci are on the line $y = k$ and are given by $(h-c,k)\,\&\,(h+c,k)$ where $c^2 = a^2 - b^2$.

   II) If $b > a$, the two foci are on the line $x = h$ and are given by $(h,k-c)\,\&\,(h,k+c)$ where $c^2 = b^2 - a^2$.

4. Hyperbola Centered at (h,k):
$$\dfrac{(x-h)^2}{a^2} - \dfrac{(y-k)^2}{b^2} = 1 \text{ or } \dfrac{(y-k)^2}{b^2} - \dfrac{(x-h)^2}{a^2} = 1$$

   I) When $\dfrac{(x-h)^2}{a^2}$ is to the left of the minus sign, the two foci are on the line $y = k$ and are given by $(h-c,k)$ & $(h+c,k)$ where $c^2 = a^2 + b^2$.

349

II) When $\dfrac{(y-k)^2}{b^2}$ is to the left of the minus sign, the two

foci are on the line $x=h$ and are given by $(h,k-c)$ &
$(h,k+c)$ where $c^2 = b^2 + a^2$

5. Parabola with Vertex at (h,k) and Focal Length p :

$$(y-k)^2 = 4p(x-h) \;\; or \;\; (x-h)^2 = 4p(y-k)$$

I) For $(y-k)^2$, the focus is $(h+p,k)$ and the directrix is
given by the line $x = h-p$.

II) For $(x-h)^2$, the focus is $(h,k+p)$ and the directrix is
given by the line $y = k-p$.

$$\int_a^b \overset{\cdot\cdot}{\smile} dx$$

Appendix C: Formulas from Finance

C.1: Simple Interest
C.2: Simple Principle Growth and Decline
C.3: Effective Interest Rates
C.4: Continuous Interest
C.5: Mortgage and Annuity Formulas

Definition of Terms

P is the amount initially borrowed or deposited.

A is the total amount gained or owed.

r is the annual interest rate.

i is the annual inflation rate.

α is an annual growth rate of voluntary contributions to a fund.

r_{eff} is the effective annual interest rate.

t is the time period in years for an investment.

T is the time period in years for a loan or annuity.

N is the number of compounding periods per year.

M is the monthly payment.

e is defined as $e = \lim_{n \to \infty}\left[1 + \frac{1}{n}\right]^n$

C.1: Simple Interest

1. Interest alone: $I = \Pr T$
2. Total repayment over T: $R = P + \Pr T = P(1 + rT)$
3. Monthly payment over T: $M = \dfrac{P(1 + rT)}{12T}$

C.2: Simple Principle Growth and Decline

1. Compounded Growth: $A = P(1 + \frac{r}{N})^{Nt}$
2. Continuous Growth: $A = Pe^{rt}$
3. Continuous Annual Inflation Rate i: $A = Pe^{-it}$

C.3: Effective Interest Rates

1. For N Compounding Periods per Year: $r_{eff} = (1 + \frac{r}{N})^N - 1$

2. For Continuous Interest: $r_{eff} = e^r - 1$

3. For Known $P, A, \& T$: $r_{eff} = \sqrt[T]{\dfrac{A}{P}} - 1$

C.4: Continuous Interest IRA

1. IRA Annual Deposit D : $A = \dfrac{D}{r}(e^{rt} - 1)$

2. IRA Annual Deposit D plus Initial Deposit P :

$$A = Pe^{rt} + \frac{D}{r}(e^{rt} - 1)$$

3. IRA Annual Deposit D plus Initial Deposit P ;
   Annual Deposit Continuously Growing via $De^{\alpha t}$:

$$A = Pe^{rt} + \frac{D}{r - \alpha}(e^{rt} - e^{\alpha t})$$

C.5: Mortgage and Annuity

1. First Month's Mortgage Interest: $I_{1st} = \dfrac{rP}{12}$

2. Monthly Mortgage/Annuity Payment: $M = \dfrac{\Pr e^{rT}}{12(e^{rT} - 1)}$

3. Total Mortgage Repayment ($P + I$): $A = \dfrac{\Pr Te^{rT}}{e^{rT} - 1}$

4. Total Mortgage Interest Repayment: $I = P\left[\dfrac{rTe^{rT}}{e^{rT} - 1} - 1\right]$

5. Continuous to Compound Interest Replacement Formula for IRAs, Mortgages, and Annuities

$$e^{rt} \Rightarrow (1+\tfrac{r}{N})^{Nt} \text{ or } e^{rT} \Rightarrow (1+\tfrac{r}{N})^{NT}$$

Note: An annuity is a mortgage in reverse where the roles of the individual and financial institution have been interchanged. All continuous-interest mortgage formulas double as continuous-interest annuity formulas.

$$\int_{a}^{b} \overset{\bullet\bullet}{\cup} dx$$

Appendix D: Summary of Calculus Formulas

D.1: Basic Differentiation Rules
D.2: Basic Antidifferentiation Rules
D.3: Lines and Approximation
D.4: Fundamental Theorem of Calculus
D.5: Geometric Integral Formulas
D.6: Select Elementary Differential Equations

D.1: Basic Differentiation Rules

1. Limit Definition of Derivative:

$$f'(x) = \lim_{h \to 0}\left[\frac{f(x+h) - f(x)}{h}\right]$$

2. Differentiation Process Indicator: $[\,]'$

3. Constant: $[k]' = 0$

4. Power: $[x^n]' = nx^{n-1}$, n can be <u>any exponent</u>

5. Coefficient: $[af(x)]' = af'(x)$

6. Sum/Difference: $[f(x) \pm g(x)]' = f'(x) \pm g'(x)$

7. Product: $[f(x)g(x)]' = f(x)g'(x) + g(x)f'(x)$

8. Quotient: $\left[\dfrac{f(x)}{g(x)}\right]' = \dfrac{g(x)f'(x) - f(x)g'(x)}{g(x)^2}$

9. Chain: $[f(g(x))]' = f'(g(x))g'(x)$

10. Inverse: $[f^{-1}(x)]' = \dfrac{1}{f'(f^{-1}(x))}$

11. Generalized Power: $\left[\{f(x)\}^n\right]' = n\{f(x)\}^{n-1}f'(x)$;
    Again, n can be <u>any exponent</u>

12. $[\ln x]' = \dfrac{1}{x}$

13. $[\log_a x]' = \dfrac{1}{x \ln a}$

14. $[e^x]' = e^x$

15. $[a^x]' = a^x \ln a$

D.2: Basic Antidifferentiation Rules

1. Antidifferentiation Process Indicator: $\int$

2. Constant: $\int k\,dx = kx + C$

3. Coefficient: $\int af(x)\,dx = a\int f(x)\,dx$

4. Power Rule for $n \ne -1$: $\int x^n\,dx = \dfrac{x^{n+1}}{n+1} + C$

5. Power Rule for $n = -1$: $\int x^{-1}\,dx = \int \dfrac{1}{x}\,dx = \ln|x| + C$

6. Sum: $\int [f(x) + g(x)]\,dx = \int f(x)\,dx + \int g(x)\,dx$

7. Difference: $\int [f(x) - g(x)]\,dx = \int f(x)\,dx - \int g(x)\,dx$

8. Parts: $\int f(x)g'(x)\,dx = f(x)g(x) - \int g(x)f'(x)\,dx$

9. Chain: $\int f'(g(x))g'(x)\,dx = f(g(x)) + C$

10. Generalized Power Rule for $n \neq -1$:

$$\int [f(x)]^n f'(x)dx = \frac{[f(x)]^{n+1}}{n+1} + C$$

Generalized Power Rule for $n = -1$:

$$\int \frac{f'(x)}{f(x)} dx = \ln|f(x)| + C, n = -1$$

General Exponential: $\int e^{f(x)} f'(x)dx = e^{f(x)} + C$

11. $\int \ln x \, dx = x \ln x - x + C$

12. $\int e^x dx = e^x + C$

13. $\int x e^x dx = (x-1)e^x + C$

14. $\int a^x dx = \frac{a^x}{\ln a} + C$

D.3: Lines and Approximation

1. Tangent Line at $(a, f(a))$: $y - f(a) = f'(a)(x - a)$

2. Normal Line at $(a, f(a))$: $y - f(a) = \frac{-1}{f'(a)}(x - a)$

3. Linear Approximation: $f(x) \cong f(a) + f'(a)(x - a)$

4. Second Order Approximation:

$$f(x) \cong f(a) + f'(a)(x - a) + \frac{f''(a)}{2}(x - a)^2$$

5. Newton's Iterative Formula: $x_{n+1} = x_n - \frac{f(x_n)}{f'(x_n)}$

$$y = f(x) \Rightarrow dy = f'(x)dx$$

6. Differential Equalities: $f(x+dx) = f(x) + f'(x)dx$

$$F(x+dx) = F(x) + f(x)dx$$

D.4: The Fundamental Theorem of Calculus

Let $\int_a^b f(x)dx$ be a definite integral representing a continuous summation process, and let $F(x)$ be such that $F'(x) = f(x)$.

Then, $\int_a^b f(x)dx$ can be evaluated by the alternative process

$$\int_a^b f(x)dx = F(x)\big|_a^b = F(b) - F(a).$$

Note: A continuous summation (or addition) process on the interval $[a,b]$ sums millions upon millions of consecutive, tiny quantities from $x = a$ to $x = b$ where each individual quantity has the general form $f(x)dx$.

D.5: Geometric Integral Formulas

1. Area Between two Curves for $f(x) \geq g(x)$ on $[a,b]$:

$$A = \int_a^b [f(x) - g(x)]dx$$

2. Area Under $f(x) \geq 0$ on $[a,b]$:

$$A = \int_a^b f(x)dx$$

3. Volume of Revolution about x Axis Using Disks:

$$V = \int_a^b \pi [f(x)]^2 \, dx$$

4. Volume of Revolution about y Axis using Shells:

$$V = \int_a^b 2\pi x \, | f(x) | \, dx$$

5. Arc Length:

$$s = \int_a^b \sqrt{1 + [f'(x)]^2} \, dx$$

6. Revolved Surface Area about x Axis:

$$SA_x = \int_a^b 2\pi \, | f(x) | \sqrt{1 + [f'(x)]^2} \, dx$$

7. Revolved Surface Area about y Axis:

$$SA_x = \int_a^b 2\pi \, | x | \sqrt{1 + [f'(x)]^2} \, dx$$

8. Total Work with Variable Force $F(x)$ on $[a,b]$:

$$W = \int_a^b F(x) \, dx$$

D.6: Select Ordinary Differential Equations (ODE)

1. First Order Linear: $\dfrac{dy}{dx} + f(x)y = g(x)$

2. Bernoulli Equation: $\dfrac{dy}{dx} = f(x)y + g(x)y^n$

3. ODE Separable if it reduces to: $g(y)dy = f(x)dx$

4. Falling Body with Drag: $-m\dfrac{dv}{dt} = -mg + kv^n$

5. Constant Rate Growth or Decay: $\dfrac{dy}{dt} = ky : y(0) = y_0$

6. Logistic Growth: $\dfrac{dy}{dt} = k(L - y)y : y(0) = y_0$

7. Continuous Principle Growth: $\dfrac{dP}{dt} = rP + c_0 : P(0) = P_0$

8. Newton's Law in One Dimension: $\dfrac{d}{dt}(mV) = \sum F$

9. Newton's Law in Three Dimensions: $\dfrac{d}{dt}(m\vec{V}) = \sum \vec{F}$

$$\int_a^b \ddot{\cup}\, dx$$

Answers to Problems

There was once a teacher of math
Who wrote limericks for a laugh.
With head in his rhyme,
He solved for a time
And reversed the rocket's path.   *July 2000*

Chapter 3: Chapter Exercises starting on page 33

1) The two large figures, where each is constructed from an identical set of four playing pieces, are not triangles. To the naked eye they appear congruent; but, in actuality, they are not.
2) The bottom Tangram figure is a square but the top Tangram figure is not a square.

Chapter 4

Section 4.1 starting on page 42

A&B)
$$(f+g)(x) = x^3 - 4x + \sqrt{x} : D(f+g) = [0, \infty)$$
$$(f-g)(x) = x^3 - 4x - \sqrt{x} : D(f+g) = [0, \infty)$$
$$(gf)(x) = \sqrt{x}(x^3 - 4x) : D(f-g) = [0, \infty)$$
$$\left(\frac{f}{g}\right)(x) = \frac{x^3 - 4x}{\sqrt{x}} : D\left(\frac{f}{g}\right) = (0, \infty)$$
$$\left(\frac{g}{f}\right)(x) = \frac{\sqrt{x}}{x^3 - 4x} : D\left(\frac{g}{f}\right) = (0,2) \cup (2, \infty)$$

C)
$$(f \circ g)(x) = \sqrt{x^3} - \sqrt{x} : D(f \circ g) = [0, \infty)$$
$$(g \circ f)(x) = \sqrt{x^3 - 4x} : D(g \circ f) = [-2,0] \cup (2, \infty)$$

D) Next page

E) $3a^2 + 3ah + h^2 - 4$

D)

Input Value	Output Value
2	0
6	192
0	0
7	315
3	15
a	$a^3 - 4a$
$a+h$	$a^3 + 3a^2h + 3ah^2 + h^3 - 4a - 4h$

Section 4.2 starting on page 46

A) $Df = (-\infty, \infty) : Dg = (-\infty, 8) \cup (8, \infty) : Dh = (-\infty, \infty)$

B) $Df = (-\infty, \infty) : Dg = (-\infty, 8) \cup (8, \infty) : Dh = [0, \infty)$

C) $f^{-1}(x) = \dfrac{11x - 5}{3} : g^{-1}(x) = \dfrac{2(1 - 4x)}{1 - x} : h^{-1}(x) = \sqrt{5x + 7}$

Section 4.3 starting on page 53

1A) 6   1B) $\frac{1}{5}$ 1C) 93   1D) 2   2) e^k 3) $6x$

4)

$pe^{rt} \Rightarrow \$5034.38$

$n = 2 \Rightarrow \$4974.47$

$n = 4 \Rightarrow \$5003.99$

$n = 6 \Rightarrow \$5014.02$

$n = 12 \Rightarrow \$5024.15$

Section 4.4 starting on page 58

1) $Df = [0, 1) \cup [2, \infty)$

2) The actual graph is left to the reader.

$$x \in (0,30] \Rightarrow f(x) = \$3.00$$
$$x \in (30,60] \Rightarrow f(x) = \$4.00$$
$$x \in (60,90] \Rightarrow f(x) = \$5.00$$
$$x \in (90,120] \Rightarrow f(x) = \$6.00$$
$$x \in (120,150] \Rightarrow f(x) = \$7.00$$
$$x \in (150,180] \Rightarrow f(x) = \$8.00$$
$$x \in (180,210] \Rightarrow f(x) = \$9.00$$
$$x \in (210,720] \Rightarrow f(x) = \$10.00$$

The function $f(x)$ is discontinuous for all x values in the set $\{30,60,90,120,150,180,210\}$. For each x in the set, f abruptly changes its functional value by $\$1.00$, creating a gap in the graph.

Section 4.5 starting on page 63

1) $m = -\frac{10}{7}; x_{int} = -\frac{39}{10}; y_{int} = -\frac{39}{7}$

2) $m = 3; x_{int} = \frac{5}{3}; y_{int} = -5$

3) $D(t) = 200 + 60t : t \in [0,5]$

Section 4.6 starting on page 71

1) The graph is left to the reader
$$g'(x) = -14x + 3; g'(-2) = 31; g'(x) = 0 \Rightarrow x = \frac{3}{14}$$

Section 4.7 starting on page 77

1) $dy = 13dx : g(x) = m = 13$ 2A) $dy = -(14x+3)dx$

2B) $dy = \dfrac{2dx}{3\sqrt[3]{(2x-5)^2}}$

Chapter 4 Review Exercise starting on page 79

$dy = (4x^3 - 4x)dx; \ f'(x) = 4x^3 - 4x$

The equation of the tangent line at $x = -2$ is $y = -24x - 40$.

Those points where the tangent line is horizontal ($f'(x) = 0$) are $x = 0, x = 1, x = -1$.

Chapter 5

Section 5.1 starting on page 87

Both $\dfrac{dy}{dx} = f(x) = -\dfrac{2}{x^3}$.

Section 5.2 starting on page 90

$$y = x \Rightarrow \frac{dy}{dx} = 1; \ y = x^2 \Rightarrow \frac{dy}{dx} = 2x;$$

$$y = x^3 \Rightarrow \frac{dy}{dx} = 3x^2; \ y = x^4 \Rightarrow \frac{dy}{dx} = 4x^3;$$

$$y = x^5 \Rightarrow \frac{dy}{dx} = 5x^4; \ y = x^{143} \Rightarrow \frac{dy}{dx} = 143x^{142}$$

Section 5.3 starting on page 107

1) $y' = 14x - 4 + 2e^x$ 2) $y' = \dfrac{2x(4x^3 + 3x^2 + x + 1)}{\sqrt{2x^2 + 1}(2x + 1)^2}$

3) $y' = x^2\left[3\ln(x^2 + 1) + \dfrac{2x^2}{x^2 + 1}\right]$

4) $y' = 12x^2 - 12$

5) $f'(x) = 4(x^4 + 12x^3 + 1)(x^4 + 1)^{11}e^{4x}$

6) $y' = \dfrac{1 - x\ln x}{xe^x}$

Section 5.4 starting on page 149

1a) Tangent line is $y = -5x + 13$

1b) Normal line is $y = \frac{1}{5}x + \frac{13}{5}$

1c) $f(\frac{3}{4}) = \frac{49}{8}$ is both a local max and global max

1d) On the interval $[-2,2]$ the absolute min is $f(-2) = -9$ and the absolute max is $f(\frac{3}{4}) = \frac{49}{8}$.

1e) On the interval $[1,3]$ the absolute max is $f(1) = 6$ and the absolute min is $f(3) = -4$.

1f) *Hint*: start with the distance formula to obtain
$$[D(x)]^2 = (x-2)^2 + (5 + 3x - 2x^2 - 1)^2.$$
Set $D'(x) = 0 \Rightarrow 4x^3 - 9x^2 - 3x + 5 = 0$ and solve the resulting polynomial equation using Newton's method.

2) The common point of tangency is $(1,3)$ and the equation of the common tangent line is $y = 2x + 1$

3a) The local max is $f(-1) = 1$. The local min is $f(-\frac{1}{3}) = \frac{23}{27}$.

3b) On the interval $[-\frac{1}{2}, 1]$, the absolute min is $f(-\frac{1}{3}) = \frac{23}{27}$ and the absolute max is $f(1) = 5$.

3c) The function f has a zero in the interval $[-2, -1]$.
Let $x_1 = -1$. The seventh iteration gives $x_8 = -1.754$, which is stable to the third decimal place with $f(x_8) = f(-1.754) = .003$.

4) Start with $V(x) = hx^2 \Rightarrow dV = 2hxdx$. For our particular set of numbers, the additional concrete needed is $66.67\,yd^3$.

5a)
$$\sqrt{76} = \sqrt{81 - 5} \cong \sqrt{81} - \frac{5}{2\sqrt{81}} = 9 - \frac{5}{18} = 8.722:$$
$$(8.722)^2 = 76.077$$

5b)
$$\sqrt[3]{29} = \sqrt[3]{27+2} \cong \sqrt[3]{27} + \frac{2}{3\sqrt[3]{(27)^2}} = 3 + \frac{2}{27} = 3.074:$$

$$(3.074)^2 = 29.050$$

6) $f(2) = 5$ is a saddle point. $f(-1) = -\frac{15}{4}$ is a local min.

7a) The maximum area is $A = \dfrac{3\sqrt{3}}{8}$.

7b) *Hint*: Use the distance formula to evaluate the distance between each pair of points in $\left\{(0,0),(1.5,0),(1.5,\frac{\sqrt{3}}{4})\right\}$.

8) For the interval $[0,10]$:

$f(0) = 0$ is an absolute min.

$f(\frac{\sqrt{2}}{2}) = .429$ is an absolute max.

$g(.5) = -.288$ is an absolute min.

$g(10) = 4.511$ is an absolute max.

9) Note that $E(x) \geq 0$ for all x. Hence, to find absolute min $E(x) = 0$, just set $E(x) = 0$ and solve for x.

10) Either $x = \sqrt{C^2 - a^2} \geq \sqrt{a(C-a)}$ or

$$x = \sqrt{2Ca - a^2} \geq \sqrt{a(C-a)}$$

11) Hard to reach hot-water pipes in a still-air space (such as a crawlspace or basement) where the temperature stays fairly uniform the year around.

Section 5.5 starting on page 159

1a) $y' = \dfrac{-(2x+3y)}{3x+2y}$

1b) $y' = \dfrac{3x^2 - 7y^2}{7(3y^2 + 2xy)}$

1c) $y' = \dfrac{1}{x+y-1}$

2) $y' = -\dfrac{2xy}{1+x^2}$, and at the point $(1,1)$ $y' = -1$.

The equation of the tangent line is $y = -x + 2$ and the equation of the normal line is $y = x$.

3) $-1.49 \frac{ft}{s}$

4) The distance between the two ships at time t is

$$D(t) = \sqrt{(40-15t)^2 + (5t)^2} . \qquad D'(t) = 0 \Rightarrow t = 2.4hr .$$

At 15:24 CST, the ships are $12.65knots$ apart.

Section 5.6 starting on page 168

To solve a grueling equation,
Rely not on the imagination.
Right answers take skill,
Much study and will,
Plus oodles of perspiration!   *July 2000*

1a) $y' = 28x^3 - 10x + 17;\ y'' = 84x^2 - 10$

1b) $y' = \dfrac{x^3 + 18x}{\sqrt{(x^2+9)^3}};\ y'' = \dfrac{162 - 11x^2}{\sqrt{(x^2+9)^5}}$

1c) $y' = (x+1)e^x - \ln x - 1;\ y'' = (x+2)e^x - \dfrac{1}{x}$

2) $y' = \dfrac{y^2 + 2x}{1 - 2xy};\ y'' = \dfrac{4y^3 - 6xy^4 + 8x^3 + 2}{[1-2xy]^3}$

3) $y' = -\dfrac{9x}{4y} \Rightarrow y' = -\frac{3}{2}$ at the point $(2,3)$.

The equation of the tangent line is $y = -\frac{3}{2}x + 6$.

The equation of the normal line is $y = \frac{2}{3}x + \frac{5}{3}$.

4) $D(t) = 100t - 10t^2 \Rightarrow D'(t) = V(t) = 100 - 20t$

   The truck stops at $t = 5\sec$ and $D(5) = 250\,ft$.

Chapter 5 Exercises starting on page 177

1) The optimum dimensions in inches are $18 \times 18 \times 36$.
   The maximum volume is $11{,}664\,in^3$.

2) Start with $S = 4\pi r^2 ; V = \frac{4}{3}\pi r^3$

   $S = 144 \Rightarrow r = 3.38in\,.\,V = 100 \Rightarrow r = 2.88in\,.$

   $\dfrac{dV}{dt} = 4\pi r^2\,\dfrac{dr}{dt} \Rightarrow 50 = 4\pi(2.88)^2\,\dfrac{dr}{dt} \Rightarrow \dfrac{dr}{dt} = .48$

   $\Delta t = \dfrac{\Delta r}{\frac{dr}{dt}} \Rightarrow \Delta t = \dfrac{.5}{.48} \Rightarrow \Delta t = 1.04s$

3a) $f(0) = 0$ is a local max. $f(\frac{2}{5}) = -.3257$ is a local min.

3b) $f(-1) = -.5$ is a local min. $f(1) = .5$ is a local min.

3c) $f(-2) = 4e^{-2}$ is a local max. $f(0) = 0$ is a *global* min.

4) $y' = \dfrac{2x - 5}{2 - 3y^2} \Rightarrow y' = 1$ at the point $(2,-1)$.

   The equation of the tangent line is $y = x - 3$.
   The equation of the normal line is $y = -x + 1$.

5) The optimum dimensions in feet are $3 \times 3 \times 3$.
   The maximum volume is $27\,ft^3$.

6) On the interval $[0,3]$, $f(0) = 0$ is the absolute max and
   $f(2) = -28$ is the absolute min.

Chapter 6

Section 6.1 starting on page 181

a) $(x+7)^2$

b) $(x+1)(x^2+3)$

c) $3(x+2)(x-2)(x^2+4)$

d) Prime

e) $2y^2(3-4y+2y^2)$

f) $(6x-5)(6x+5)$

g) $(3x-1)(2x-1)$

h) $(x-4)(x+3)$

i) $2x(4x-1)(x+3)$

j) $(3x+2)(x-4)$

k) $3(m-5n)(m+2n)$

l) $(5x+2)^2$

Section 6.3 starting on page 205

1) $h(t) = 2 - \dfrac{\sqrt{(4-t^2)^3}}{3}$

2) $G(s) = (s^3 - 3s^2 + 6s - 6)e^s + 7$

3) $P(x) = \dfrac{x^3}{3} + 2x^2 - 5x + \dfrac{20}{3}$

4) See the **Stern Warning** on page 201

Note: All checks are left to the reader in problem 5.

5a) $\dfrac{3\sqrt[3]{x^4}}{4} + C$

5b) $\dfrac{x^4}{4} + x^3 - \dfrac{x^2}{2} + C$

5c) $\dfrac{x^4}{4} + x^3 + \dfrac{1}{x} + C$

5d) $\dfrac{5x^8}{2} + 8x^5 + 10x^2 + C$

5e) $\dfrac{x^2}{2} + \dfrac{4\sqrt{x^3}}{3} + x + C$

5f) $\dfrac{5x^2}{2} + 10x + C$

5g) $\dfrac{x^2}{2} + \dfrac{x^3}{3} + \dfrac{x^4}{4} + \dfrac{x^5}{5} + \dfrac{x^6}{6} + C$

5i) $t\ln t - t + C$

5j) $\dfrac{x^3}{3} + \dfrac{3x^2}{2} + 2x + C$

5k) $\dfrac{e^{5x}}{5} + \dfrac{e^{4x}}{2} + \dfrac{e^{3x}}{3} + C$

5l) $\dfrac{2(1+\sqrt{x})^5}{5}+C$

5m) $\dfrac{(2a^2+3)^{1002}}{4008}+C$

5n) $\dfrac{-e^{\frac{10}{x}}}{10}+C$

5o) $\dfrac{2(w^3+1)^{13}}{39}+C$

Chapter 6 Chapter Exercises starting on page 221

1a) $y(x)=\dfrac{2}{\sqrt[4]{1-64x-32x^2}}$ 1b) $y(x)=\sqrt{\ln\left[\dfrac{x(x+1)}{2}\right]+1}$

2) Start with $D''=-32$ to obtain $D(t)=1450+50t-16t^2$. From this equation, we can determine all subsequent quantities.

> Time to impact: $11.21\sec$
> Impact velocity: $-308.71\frac{ft}{\sec}$ or $-210.48mph$
> Time to apex: $1.5625\sec$
> Maximum height or apex: $1489.06\,ft$ above ground

3) Newton's Law of Cooling for this particular set of conditions is $T_B(t)=34+98.6e^{-1.2416t}$. The expression $T_B(t)$ is bulk body temperature as a function of time. Setting $t=1hr$ gives $T_B(1)=62.48^0F$. Unfortunately, this is below the critical temperature of 65^0F. However, Newton's model is a crude estimate that assumes temperature uniformity throughout the body. This is definitely not the case with the human body, which—in a condition of rapid cooling—shuts down blood flow to the extremities in order to keep the vital organs in the interior as warm as possible. Hence, our victim still has a chance—but not for long. Our model adds urgency to the rescue attempt!

Chapter 7

Some areas are hard to calculate
Inspiring some wits to speculate.
They fiddle and horse
And finally curse
Because they refuse to integrate!   *July 2000*

Section 7.1 starting on page 229

1a) $A = \left[\dfrac{x^5}{5} + \dfrac{x^3}{3} \right] \Big|_0^4 = \dfrac{3392}{15}$

1b) $A = \left[\dfrac{x^5}{5} + \dfrac{x^3}{3} \right] \Big|_{-4}^4 = 2 \left[\dfrac{3392}{15} \right] = \dfrac{6784}{15}$

2)

$\overset{1}{\mapsto} : A(z) = \left[\dfrac{x^3}{3} + 2x \right] \Big|_0^z = \dfrac{z^3}{3} + 2z - 5$

$\overset{2}{\mapsto} : A(z) = 10 \Rightarrow z^3 + 6z - 75 = 0 \Rightarrow z \cong 3.74495$

3) Let $x_1 \geq 0$ be such that $0 < x_1 + b < B$. Then the total area of the trapezoid is given by the three *definite integrals*

$A = \int_0^{x_1} \left[\dfrac{hx}{x_1} \right] dx + \int_{x_1}^{x_1+b} h\, dx + \int_{x_1+b}^{B} \left[\dfrac{B-x}{B-x_1-b} \right] dx \Rightarrow$

$A = \dfrac{hx_1}{2} + hb + \dfrac{h}{2}[B - (x_1 + b)] = \dfrac{(b+B)h}{2}$

Section 7.3 starting on page 238

1a) $\dfrac{13}{3}$

1b) $\dfrac{2296}{3}$

1c) $\dfrac{[(\ln 4 + 1)^3 - 1]}{3} = 4.196172$

1d) $\dfrac{88{,}573}{11}$

2) $A = \int\limits_{1}^{4}(x+\sqrt{x})dx = \dfrac{73}{6}$

Chapter 7 Chapter/Section Exercises starting on page 262

1)

$$S = \pi a^2 + \pi b^2 + \int\limits_{0}^{h}\left[2\pi\left\{a+\dfrac{(b-a)x}{h}\right\}\sqrt{1+\left\{\dfrac{b-a}{h}\right\}^2} \right]dx \Rightarrow$$

$$S = \pi(a^2+b^2)+2\pi\left[\dfrac{a+b}{2}\right]\sqrt{h^2+(b-a)^2}$$

2) $V = \int\limits_{0}^{h}\pi\left[a+\dfrac{(b-a)x}{h}\right]^2 dx = \dfrac{h(b^2+ab+a^2)\pi}{3}$

3) $A = \int\limits_{1}^{3}\left[4-x-\{x^2-5x+7\}\right]dx = \dfrac{4}{3}$

4)

$\overset{1}{\mapsto}:s = \int\limits_{1}^{3}\sqrt{1+7^2}\,dx = 10\sqrt{2}$

$\overset{2}{\mapsto}:D = \sqrt{(3-1)^2+(22-8)^2} = 10\sqrt{2}$

5)

$\overset{1}{\mapsto}:S = 2\int\limits_{0}^{r}2\pi\sqrt{r^2-x^2}\left[\sqrt{1+\left\{\dfrac{-x}{\sqrt{r^2-x^2}}\right\}^2}\right]dx \Rightarrow$

$S = 4\pi r^2$

$\overset{2}{\mapsto}:V = 2\int\limits_{0}^{r}\pi\left[\sqrt{r^2-x^2}\right]^2 dx = \dfrac{4\pi r^3}{3}$

6A) $A = \int\limits_{2}^{3} x^3 \, dx = \dfrac{65}{4}$

6B) $V_x = \int\limits_{2}^{3} \pi [x^3]^2 \, dx = \dfrac{2059\pi}{7}$

6C) $V_y = \int\limits_{2}^{3} \pi x [x^3] \, dx = \dfrac{422\pi}{5}$

6D) $V_{y=-2} = \int\limits_{2}^{3} \pi [2 + x^3]^2 \, dx = \dfrac{2542\pi}{7}$

6E) $V_{x=-2} = \int\limits_{2}^{3} \pi (x+2)[x^3] \, dx = \dfrac{747\pi}{5}$

Chapter 8

Section 8.1 Exercises starting on page 265

1a) Implicit: $y(x) = \dfrac{6}{5 - 2x^3}$

1b) Explicit: $y(x) = \dfrac{x^3}{3} + \dfrac{x^2}{2} + 3$

1c) Implicit: $y(x) = \sqrt{\ln\left[\dfrac{x-2}{2}\right]^2 + 1}$

1d) Explicit: $y(x) = e^{4x}$

2) $A(x) = \dfrac{35x^5 + 3840}{2480} \Rightarrow A(6) = \dfrac{3450}{31}$

Section 8.2 Exercises starting on page 292

1) Apply $V_{sun-surface-escape} = \sqrt{2g_{sun}R_{sun}}$ to obtain $g_{sun} = 901\frac{ft}{s^2}$ at the sun's surface. The escape velocity at the point of the earth's orbit is given by the expression

$$V_{earth-orbit-escape} = \sqrt{2g_{sun}\left(\frac{R_{sun}}{R_{earth-orbit}}\right)R_{sun}} = 26.16\frac{miles}{s}$$

2) $v(t) = \dfrac{(46-4600t)^2}{4}$, $k = 66.24$, and the impact force is

$-k\sqrt{v(0)} = 1523.52\,Newtons$.

$$D(t) = 1.76333 - \frac{(46-4600t)^3}{55,200} \Rightarrow D(.01) = 1.76333m$$

3a) $y(x) = \dfrac{3e^{2x}-1}{2}$

3b) $y(x) = \dfrac{2e^{2x}}{3-e^{2x}}$

3c) $y(x) = 2\sqrt{\dfrac{e^{4x}}{6-2e^{4x}}}$

3d) See **Ex 8.15**, page 300

4)

$$y(t) = \frac{40,000}{50 + 750e^{-0.439445t}} ;$$

$$75\% \Rightarrow y(t) = 600 \Rightarrow t = 8.67\,years$$

5) $W = 9.6 \displaystyle\int_{10}^{15}(x-5)dx = 360in \cdot lbf$

6)

$$L_{\%}(t) = e^{-0.069315t} \Rightarrow$$

$$L_{\%}(43.2 \, years) = .05 \,\&\, L_{\%}(99.66 \, years) = .001$$

7) $404.03 \, ft$

8) $W = \int_{0}^{10}(x^2 - 10x)dx = 166.67$ units of work

$F(5) = 25$ units of force

Section 8.3 Exercises starting on page 307

1)

Fixed Rate Mortgage with $P_0 = \$230,000.00$				
Terms	r	M	A	A_{PV}
$T = 30$	6.50%	$1452.49	$522,894.30	$344,779.27
$T = 20$	6.25%	$1678.94	$402,945.95	$303,007.54
$T = 15$	5.50%	$1876.52	$337,774.69	$272,000.08

2A) 17.33 years   2B) 10.98%   2C) 7.49%

3A) $6,272,371.03 face value

3B) $1,675,571.72 present value

3C) $7,859.72 present value of first monthly annuity payment

3D) $3,101.08 present value of last monthly annuity payment

Short Bibliography

History of Mathematics

1. Ball, W. W. Rouse; <u>A Short Account of the History of Mathematics</u>; Macmillan &co. LTD., 1912; Reprinted by Sterling Publishing Company, Inc.,2001.

2. Beckman, Peter; <u>A History of PI</u>; The Golem Press, 1971; Reprinted by Barnes & Noble, Inc., 1993.

General Mathematics

3. Hogben, Lancelot; <u>Mathematics for the Million</u>; W. W. Norton & Company, 1993 Paperback Edition.

Introduction to Calculus

4. Thompson, Silvanus P.; <u>Calculus Made Easy</u>; The Macmillan Company, New York, 1914; Reprinted by St. Martin's Press in 1998 with Additions and Commentary by Martin Gardner.

5. Silverman, Richard A.; <u>Essential Calculus with Applications</u>; W.B. Saunders Company, Philadelphia, 1977; Reprinted by Dover Publications, 1989.

Standard College Calculus

6. Stewart, James; <u>Calculus 4th Edition</u>; Brooks/Cole Publication Company, 1999.

7. Thomas, George B. Jr. & Finney, Ross L.; <u>Calculus and Analytic Geometry 8th Edition</u>; Addison-Wesley Publishing Company,1992.

8. Fobes, Melcher P. & Smyth, Ruth B.; <u>Calculus and Analytic Geometry, Volumes I & II</u>; Prentice Hall Inc., 1963; *Out of Print*.

Index

Differential Equation